The World System

THE INTERNATIONAL LIBRARY OF SYSTEMS THEORY AND PHILOSOPHY
Edited by Ervin Laszlo

THE SYSTEMS VIEW OF THE WORLD
*The Natural Philosophy of the New
Developments in the Sciences*
by Ervin Laszlo

GENERAL SYSTEM THEORY
Foundations—Development—Applications
by Ludwig von Bertalanffy

ROBOTS, MEN AND MINDS
Psychology in the Modern World
by Ludwig von Bertalanffy

THE RELEVANCE OF GENERAL SYSTEMS THEORY
*Papers Presented to Ludwig von Bertalanffy
on His Seventieth Birthday*
Edited by Ervin Laszlo

HIERARCHY THEORY
The Challenge of Complex Systems
Edited by H. H. Pattee

THE WORLD SYSTEM
Models, Norms, Applications
Edited by Ervin Laszlo

Further volumes in preparation

THE WORLD SYSTEM

Models, Norms, Applications

Edited by ERVIN LASZLO

GEORGE BRAZILLER   New York

ACKNOWLEDGMENT

The editor and publisher wish to thank the following for permission to reprint Figures 1–4 in the essay titled, "Some Political Implications of the Forrester Model" by Alastair M. Taylor:

Prentice-Hall for Figure 1, from *A Framework for Political Analysis,* by David Easton © 1965.

Alfred A. Knopf, Inc. for Figure 2, from *Politics Among Nations,* by Hans Morgenthau © 1964, Third Edition.

John Wiley and Sons, Inc. for Figures 3 and 4, from *Systematic Political Geography,* by H. J. de Blij © 1967.

For information address the publisher:
George Braziller, Inc.
One Park Avenue, New York, N.Y. 10016

Standard Book Number: 0-8076-0695-2, cloth
                      0-8076-0696-0, paper

Library of Congress Catalog Card Number: 73-79050

First Printing
Printed in the United States of America

Preface

IN THE remaining decades of this century, mankind's problems will be increasingly complex in detail and global in scope. They will also be increasingly critical for human survival and civilization. Never before have so many people faced so many problems of such great complexity. Any attempt to isolate issues and apply short-range remedies will continue to fail by reason of the growing interdependency of all vital processes on this planet.

General systems theory is the branch of science specifically designed to cope with complexity. It makes sense to attack our present and future problems through concepts and principles developed in this theory. And it is not accidental that more scientists and humanists are now turning to systems theory for solution than ever before: whenever problems are due to interrelated processes on multiple levels, the systems approach has a selective advantage over all others.

"World system" is the conceptualization fittest to handle mankind's current needs. Different contents can be assigned to this concept, but its basic scope and nature is clear: the world system is that sphere of multilevel interdependencies which unites the planet's human population with its organic and inorganic environment. The future of this system is at stake; and the resolution of the problems threatening it determines the fate of mankind, its culture and civilization.

World system modeling is a new art. It makes use of scientific data, system dynamics principles, and computer-simulated projections. In its present stage of development (represented by the work of Forrester, Meadows, and collaborators), it raises a multitude of fundamental issues. Foremost among these are questions concerning the completeness of the variables, the accuracy of the represented system dynamics, the incorporation of normative or "soft" data, the evaluation of the findings, the uses and misuses of existing models, the conceptual and behavioral reorientation presupposed or suggested by the models, and the understanding of the methods and principles by which they can be further refined.

These are among the issues considered by contributors to the present volume. More specifically, Margaret Mead and Ervin Laszlo debate the practical use and effect of existing models; Alastair Taylor and Richard Falk explicate their implications for political thought and action; Henryk Skolimowski and Albert Wilson investigate the presuppositions of world system theories and trace the shift in scientific modes of thinking: Håkan Törnebohm outlines the structure of inquiry in studying research itself, i.e., "studies of studies"; and Ralph Burhoe elucidates the roles and functions of human values in the world system and calls attention to their isomorphy with the concepts and precepts of traditional religions. Jay Forrester assesses the problems and potentials of world system studies and replies to the main lines of criticism.

The papers fall into two broad classes: one class moves on the interface between theory and practice and concerns itself with defining the nature of the desired models and the range of their applications. Papers in this class comprise the material of Part One. The other class penetrates to the interface between theory and metatheory, examining the norms, methods, and presuppositions which guide world systems model-building. Papers in that class are grouped in Part Two.

A significantly broad range of issues of current relevance is discussed here, from differing perspectives and often with divergent consequences. But the discussions are united in their diversity through the common language of general systems theory and the common aim of developing a humanistic body of scientific knowledge. These are basic characteristics of systems philosophy, as it informs the thinking of the writer and the spirit of the International Library of Systems Theory and Philosophy. This volume resulted from the first of an annual series of Systems Philosophy Symposia,* devoted to the multidisciplinary discussion, through general systems theory, of current topics of human and philosophical interest.

E. L.

The Center of International Studies,
Princeton University

---

* Held September 29 and 30, 1973, at the State University of New York at Geneseo.

Contents

PART TWO
THEORY AND METATHEORY:
NORMS, METHODS, AND PRESUPPOSITIONS

PART ONE

THEORY AND PRACTICE:

MODELS AND

APPLICATIONS

1

Uses and Misuses of World System Models

ERVIN LASZLO

CONCERN and controversy over the state of the world is currently growing at an exponential rate. Debate and discussion focuses on the question whether the world system, composed of the human population of the earth together with its technology and life-supporting ecology, can tolerate further growth without limit, or when and how limits to growth must be introduced.

It has become evident, I believe, that a sophisticated model of the world system is needed if mankind's response to what may be an impending crisis is to be effective. It has also become evident, at least to me, that such a model must be a systems model, capable of mapping an entire array of interacting variables and their effect on the state of the world over time. We do have some systems models of global scope, and our purpose in this volume is to discuss their cogency, effectiveness, and readiness for implementation.

The argument I am advancing is that no model in existence today is sufficiently complete and free from error to warrant implementation on a global scale. On the other hand to sit back and wait until sufficiently sophisticated models are developed may be lethal: by that time world systemic processes may have shifted to catastrophic pathways. If my argument is correct, it is improper to act now, and folly to wait until later. Faced with this dilemma, I shall attempt to outline an alternative course, which combines responsible action with a high degree of independence from possible fallacies in current models. But first I shall discuss why I believe that current world system models are not evolved enough to be reasonably implemented on a global scale. I choose the most discussed and explicit of existing models for my discussion: that of Forrester.[1]

I

Forrester's world system model is based on selected empirical information, and the computer runs on the model in which the information has been integrated. It is not an intuitive model; in fact many of its findings appear directly counterintuitive. Experience with smaller scale models has shown, however, that intuitive mental models do not stand the test of time, inasmuch as the expected and the actual consequences tend to conflict. The Forrester model (as well as those which follow upon it in the work of Dennis Meadows and his collaborators[2]) is not exposed to this danger: the initial values for the selected variables are fed into a computer and the consequences read off the printout. However, a computer printout is only as good as the program on which it is based, and the program is based on certain assumptions about a territory that is mapped. These assumptions must be adequate if the computer runs are to give projections which are sound enough to justify global implementation. My first objective is to point out that in some important respects the assumptions underlying the Forrester model are not adequate to that degree.

Before making good this claim, I should point out that neither Forrester nor his associates regard their world system model as a blueprint for global reform. They are among the first to admit the preliminary and tentative nature of their findings. Nevertheless, Forrester, Meadows, et al., do sound urgent warnings and they do wish to be taken seriously. And these warnings flow out of the consequences of the computer projections based on their model. Hence there is a thin line between the cautiousness of a sincere scientist and the concern of a genuine humanist. One purpose of my critique is to resolve the existing tension between these factors and render them compatible—indeed, mutually reinforcing.

First, let me note that the Forrester model distributes the values of its principal variables homogeneously over the globe, and disregards thereby the specifics of existing social and ecological systems. For example, much of pollution is local or regional in its effects; population growth differs in local areas by as much as a factor of six; industrialization and food production are likewise highly divergent in different regions. Yet in the Forrester model the world system forms one vast and homogeneous social system, embedded in an equally vast and homogeneous natural system. The homogenization of real heterogeneity has untoward consequences when it comes to prompting changes in actual policy making (cf. Mead, below), and it signifies a flaw in the model and entail inaccuracies in its deduced consequences.

Second, the variables entering the Forrester model are true but incomplete. Those that enter into the model are all economic and ecological factors, whereas a number of psychological, social, political, moral, and cultural variables also need to be taken into account. Again, it must be pointed out that Forrester is well aware of this. He does say of his own model that "it focuses on a few major factors and omits most of the substructure of world social and economic activity."[3] And he does attempt to account for at least some of these additional factors when he speaks of ethical values as components of the world system, and the long-range goals of a society as major factors in its survival.[4] However, despite recognizing the simplifications of the existing model, Forrester appears to regard it sufficiently complete to base his perspectives of the future on it. His perspectives define a basically expansionist view of the human component of the world system, and no self-restraining adaptive capacities. Would the additional factors recognized by Forrester actually enter into his world system model, they might change its behavior sufficiently for these views to be deprived of foundation. I now turn to an analysis of the model to elucidate this possibility.

The model includes five "system levels" which function as its key variables. Each level represents one type of accumulation or integration, having input and output flows with specific rates. The levels accumulate a net quantity that results from the flow rates that add to and subtract from their values. Forrester distinguishes the following levels: population, capital investment, natural resources, fraction of capital devoted to agriculture, and pollution. His assumptions, relevant to each level and their flow rates, are as follows:

Population: An increase in population causes an increase in birth rate, since birth rate is calculated as a percentage of the existing total population. Higher birthrate means further increased population, which in turn adds to the increase of the birth rate. At the same time an increase in population triggers an increase in the death rate (if there are more people, more people die in absolute numbers) and this decreases the population. Population is constant only when birth and death rates are in equilibrium.

Rephrased, the assumption is that people tend to have a constant number of children; consequently if there are more people, there are more children; and when the children grow up they too will have about as many children as their parents. Provided that more people come into the world than die, population increases.

Capital investment: Capital-investment generation depends on population and on a normal rate determining the capital units per person per year that will be generated under normal conditions. At low values of material standard of living, pressures to consume all output are so great that little capital can

6

be accumulated. The ability to create new capital increases as the amount of capital per person increases. But at very high levels of capital investment per person, the need and incentive for further increasing the material standard of living decreases, and the generation of new capital no longer rises with higher levels of capital.

Rephrased, the assumption is that people tend to invest in capital proportionately to their standard of living; people who have little invest little and consume all that is available to them; those who have more invest more. The more people invest in capital the better are the material conditions for earnings—hence people have more money to invest. But when people have more money than they can spend on material comforts, incentives for further investments decline.

Pollution: Pollution is dissipated in natural processes below a given threshold. If pollution increases within that threshold, pollution absorption increases to decrease pollution. The threshold of absorption correlates with a time-constant: when there is little pollution, it can be dissipated quickly, but more pollution takes longer to absorb. Thus if pollution increases, pollution-absorption time increases, reducing the rate of pollution absorption and increasing the rate of pollution.

In plain language the assumption is that if man creates pollution faster than nature can absorb it, he will have progressively more pollution.

Population controlled by crowding: If within a geographic area of constant size population increases, the crowding ratio increases. This causes the birth rate to decline and brings about a reduction of the population. Likewise if crowding ratios increase, the death rate rises and the population is decreased. The double effect of reducing the birth rate and increasing the death rate acts as a powerful stabilizer of population at the maximum tolerable value of crowding.

The assumption here is that if men are too tightly crowded

they have fewer children and more deaths. Hence population thins out until crowding becomes tolerable.

Population controlled by food supply: If population rises, the ratio of invested capital per person decreases, together with the fraction of capital invested in agriculture. Consequently food potential per person decreases, bringing about a corresponding decrease in birth rate and therewith in population. The decrease in food potential per person brings about a corresponding increase in the death rate, which further decreases the population. Hence population stabilizes at the verge of starvation.

This particular assumption states simply that people have as many children and as few deaths as their food supply allows.

Population controlled by pollution: An increase in population increases pollution generation, thus increasing the total level of pollution as well as the ratio of pollution to people. The increased pollution reduces the birth rate and increases the death rate to reduce population. At the same time the rising pollution ratio reduces the food supply and therewith further reduces the birth rate and increases the death rate.

Restated, this means that if there is more pollution, people live less healthily and have less food. Thus fewer of them get born and more of them die.

Population controlled by natural resources: If population increases, the natural-resource-usage rate increases and speeds the depletion of natural resources. Thus the fraction of natural resources remaining is lowered together with the ability to extract further natural resources. This in turn reduces the effective capital-investment ratio per person and herewith the material standard of living. The falling standard of living increases the birth rate, but increases even more the death rate. The total population is reduced, and a more favorable ratio established for the natural-resource-usage rate.

The thrust of this assumption is that if you have an increasing number of people using a constant nonrenewable supply of resources, sooner or later their material standards of life are going to suffer. Then they may have more children but will have even more deaths, until there are few enough left to make do with the reduced supply of resources.

I have given a summary of Forrester's formulations together with their paraphrasing in plain language. But to get to the essence of the assumptions, one can perform a still further reduction. In this formulation we get seven basic assumptions adding up to a single master assumption.

1. The more people, the more children—the more children, the more people.
2. The better off people are, the better off they will be thereafter (until some upper limit is reached).
3. If people pollute more than nature can take, people get more pollution.
4. There tend to be as many people as there is room for.
5. There tend to be as many people as there is food for.
6. Pollution kills people and reduces their food supply.
7. There tend to be as many people as there are natural resources for.

These assumptions yield the general proposition: *There tend to be as many people as there are utilized natural resources and unpolluted food and room for.*

The thrust of the conclusion is not accidental: the very structure of the principles underlying the model presupposes it. The negative-feedback loops operate only when critical thresholds are reached—for example, when crowding reaches the maximum density that man can tolerate psychologically, or when population reaches the maximum number that can be fed, housed, and that can survive their own pollution. The basic assumptions are clearly expansionist. Only four factors capable of limiting the expansion of human population and its technological drive are operative: depletion of natural resources, rise of population, increase in crowding, and decline in food supply. Since all moral, psychological, political, and cultural self-regulations of human populations and technologies are filtered out of the model, the choice it offers contemporary mankind is simple and drastic: we either introduce voluntary pressures *now*, or the system will introduce them later by its own dynamics. The latter alternative is likely to be unpleasant, involving a decimation of the human population by starvation, overpollution, overcrowding, falling material standards of living, malnutrition, and attendant social and physical ills. So long as supraeconomic and ecological factors are recognized in the abstract but not integrated in the model which serves as the basis for our perspectives of the future, it is natural to assume that population grows until it encounters limiting pressures in the form of drastic reductions in the material quality of life. So long as the interaction of value dynamics embodied in individual as well as social and institutional behavior is not mapped in the model, the only alternative it permits is to introduce pressures *now*, rather than wait for their occurrence later. The pressures to be introduced now include, in the light of the Forrester projections, immediate reductions of the natural-resource-usage rate by 75 per cent, pollution generation by 50 per cent, capital-investment generation by 40 per cent, food production by 20 per cent, and birth rate by 30 per cent.

The alternative is no less drastic in the related Meadows model. According to these projections, the world system can be stabilized only if (1) the birth rate is made to equal the death rate by 1975, (2) the investment rate equals the depreciation rate by 1990, (3) the composition of the GNP is shifted away from material products towards services, (4) industrial capital and material goods are redesigned to prolong their usefulness, and (5) investment programs are shifted toward conservation, soil enrichment, and increased food output regardless of existing demand.[5]

An unfortunate disregard for the adaptive dynamics of supraeconomic and ecological factors led a team of British ecologists to suggest even more drastic and specific steps to ensure the survival of humanity. Their "Blueprint for Survival" proposes a global strategy of controlling world system variables, starting with the establishment of national population services, the introduction of taxes on raw materials, amortization, and power, and ending with the decentralization of industry, the redistribution of government, and the constitution of self-sufficient communities of about 500 persons each, spread over the habitable globe.[6]

Because of the unilateral focus of existing world system models on basic economic-ecological factors, we obtain an oversimplified, and (hopefully) unnecessarily pessimistic, perspective on the future. Consequently the dilemma posed by them in the eyes of sincere humanists may likewise be spurious. The only alternative to foolish trepidation or optimistic postponement of action may not be decisive (and possibly misguided) action *now*. The alternative may be the mobilization of adaptive capacities in the supraeconomic and ecological dimensions in the world system by purposive, yet not coercive, policies. It is this alternative that I wish to explore next.

I I

I suggest that the moral, psychological, and cultural aspects of existing multiperson systems possess pattern maintaining and evolving adaptive mechanisms which can be tapped for guiding the transition of the world system from runaway growth to dynamic steady state. The possibility of easing the transition and channeling its thrust in this way needs careful consideration. I offer here two basic principles which may serve as working hypotheses of research in this area.

Detailed planification of the states of complex open systems is not necessary. Complex open systems which arise in the biosphere independently of conscious human planning include biological, social, cultural, as well as ecological systems. These systems have pattern-maintaining as well as pattern-evolving dynamics built into them without benefit of human engineering. In such dynamically stable systems deviation by any part is compensated by the corrective behavior of the rest. The number of deviations from the steady-state norms can be very large, and the repertory of responses from which each part would have to select the one that compensates for the deviation would be almost infinite. The proper response by the part would have to take into account the normative steady state (or growth pattern) of the system; the changes introduced into these states by the deviating components; and the effect of its own behavior on the rest of the system. The calculation needed to come up with the indicated response pattern in a system no larger than a flea calls for equipping each of its cells with a megacomputer, and the rationality to use it. Obviously, such techniques are not used by existing natural, social, and cultural systems.

Just what the technique is that such systems use to assure the cooperation of their parts is not fully known. Paul Weiss suggests that we may need to replace concepts such as "cooperation of parts" with concepts of continuous fields.[7] The field continua may have preferred configurations to which they return when disturbed, and the entities we identify as parts may be focal points or condensations of field strength along continuous but complex matrices. Be that as it may, it is certain that steady states in systems are not simple states of mechanical equilibrium, and they are not engineered by the parts through conscious reasoning. Inasmuch as the world system was not consciously planned, but arose out of the welter of interacting ecological, economic, social, and cultural factors at work in the world, it too, is a variety of complex open system. To render it dynamically stable, it is sufficient to establish a high level of purposive information-flow among its human members. This leads to my second working hypothesis.

People must be sensitized to world system constraints on their behavior through effective networks of public information. In natural systems below the level of multihuman systems, the parts are continuously and effectively intercommunicating. Each part is sensitive to systemic constraints which limit its intrinsic degrees of freedom. Exceptions are striking and usually lethal to the system: cancer is a case in point. However, the effectiveness of human communication systems can be easily impaired; individual human beings, and their primary social, economic, and political systems, can become insensitive to world systemic constraints. Until recently, the earth was large enough, and its resources vast enough, to make the effects of such lack of sensitivity negligible. This is no longer the case. Thus we must resensitize each other to world systemic constraints, and this can only be done through effective intercommunication. Such communication can revive latent adaptive mechanisms in people, in social groups, and in cultures as a whole. If successful, the result of effective intercommunication is the spontaneous adaptation of human, national, and supranational goals to the objective requirements of a global steady state.

The program of public information and education I have in mind would draw on existing world system models as its primary source of information. But it would not commit the mistake of taking any existing model as a blueprint for global action, for the sake of saving mankind from its own follies.

Rather, it would use world system models as means to inform and sensitize people concerning the global consequences of their behavior. Such use has in fact been foreseen even for the Forrester model, notwithstanding its conclusions calling for immediate and effective pressures. Aurelio Peccei said that the models should be used "to move men on the planet out of their ingrained habits." If a value reorientation is to be brought about, men must be alerted to the fantastic problems they collectively face in the next few decades; they must be sensitized to the dangers attendant on overstepping the limits of steady states in the world system. Hence there is an urgent need for a warning signal in regard to every inherently dangerous trend in public behavior. World system modeling can provide that warning signal. It can guide mankind along a pathway of damped deviation from the narrow road of biocultural survival, as an automatic pilot guides a plane along its glide path, and as a conscience guides a moral person along the ethical pathways of action.

People are, in fact, interested enough in the long-range destiny of mankind to do something about it, even if it goes against their current short-term interests. Forrester found such willingness in his work with urban systems, and the world-wide interest accorded the Club of Rome projections indicates that people are indeed deeply concerned with the extended time-horizon beyond tomorrow.

We seem to be moving on the right track. Few world and government leaders seem ready to enforce vast programs to produce the pressures needed to conform to the existing projections for an equilibrium world system. On the other hand, world leaders, as well as the man on the street, show concern and willingness to listen. What we tell them is of paramount importance. It can be schematically divided into two types of signals: *warning signals* and *suggestions*.

(1) Warning signals are already produced by a steadily

14

increasing number of intellectuals—scientists and humanists. They confront people with the consequences of persisting in behavior patterns that conduce to unlimited growth in the world system. The present concern over pollution, population, use of natural resources, availability of food supplies, international strife and war, are among the results of warning signals heard today in many parts of the globe. The publicity accorded the Club of Rome projections is a good instance of the effectiveness of warning signals to command public attention. From public attention and genuine concern to modified behavior is another step, but one which is not impossible to take. If people are sufficiently concerned with, and aware of, problems in their daily practice of life, the inertia of habit and cultural conditioning is already halfway overcome. New generations are less subject to conditioning for the status quo, and the effect of warning signals on them can be a welcome liberation from established values and attendant patterns of behavior. This takes me to the second type of signal: *suggestions*.

(2) The effect of warning signals is to loosen ingrained habits and create an openness for new values and actions. But what values and which actions? Answers to such questions may emerge in profusion when awareness has reached a sufficient level, and people are acutely disturbed by the insufficiency of existing modes of behavior. Yet it is risky to adopt a laissez-faire attitude in the comforting belief that new and adequate values will emerge spontaneously when the time is ripe. When time thresholds are critical, lags between real conditions and their perception can cause delays compounded by the lag attendant on the resistance of traditional values and action patterns to innovation. Even when openness to change is purposively created, residual time lags may exceed critical thresholds and impel the world system toward catastrophe. *Hence simulation in the area of new values and their inter-*

15

action with basic economic-ecological processes is needed. World system models must include system levels in the form of social, moral, and cultural values, interacting with the basic levels of food production, pollution, industrialization, natural-resource usage, and population. Such integrated models will suggest viable constellations of values, just as they now suggest viable trends in the basic economy and ecology. The viable values can be the content of the suggestions emanating from world system research and giving positive direction to the relatively unconstrained dynamics of goals and purposes created by a consistent use of the warning signals.

At the Woodrow Wilson Symposium in Washington, on March 2, 1972, Senator Pell (D., Rhode Island) said to Dennis Meadows, "You presume man is rational, but in our work he is emotional. How do you convert this into an action program?" The answer, I suggest, must include the creation of an institutional mechanism for continuously and effectively putting out two kinds of signals for feedback–guiding the processes of transition to steady state.[8] The warning signals of computer simulations of catastrophic pathways of development give the sharp beep-beep of deviation from the feasible glide path. The positive signals of recommended modes of value and action patterns provide the reassuring continuous tone of being on target. The spontaneity of personal, organizational, national, and supranational motivation can then do the actual work of implementing the transition.

CONCLUSIONS

I have attempted to point out the principal uses and misuses of world system models. I found that the principal use of such models is to provide information leading to the voluntary reorganization of existing social behavior patterns through the internalization of more adequate value systems. The principal and potentially dangerous misuse of world system models is to attempt to implement their findings as blueprints for global reform. Since current models tend to concentrate on basic biologic, economic, and ecological factors, disregarding moral, psychological, and cultural dimensions, global reform blue-printed by existing models may turn out to be highly coercive. Fortunately it may not be necessary: the much discussed but so far not effectively mapped moral, psychological, and cultural dimensions of the world system may incorporate suffi-cient reserves of adaptive potential to permit the transition from unlimited growth to dynamic equilibrium by the reorien-tation of people's thinking rather than by pressures acting against their perceived interests.

Thomas Jefferson said, "I know of no safe depository of the ultimate powers of society but the people themselves; and if we think them not enlightened enough to exercise their control with a wholesome discretion, the remedy is not to take it from them, but to inform their discretion." Hence the way out of the double-bind of whether to create possibly arbitrary pressures now, or to wait helplessly and hope that the prob-lems resolve themselves, is to "inform people's discretion." As long as enlightened information can be made available, it remains the clearly preferable alternative to the use of force as well as to the posture of wait and see.

2

Models and Systems Analyses as Metacommunication

As a Society* devoted to the development of systems analysis, I believe that it is one of our responsibilities to be alert to the kinds of metacommunication, communication *about* systems analysis, which results from the use of particular models or the use of models for particular purposes. Thus the use of a *world* model may convey the idea, which is not yet generally understood, that the world is indeed an interconnected system and that events in one part of the world are significant just because they will have repercussions in another part of the world. It may also convey the idea that the world is a closed system in which ideas of confinement, limits, and danger may accompany the use of the model. Man, proclaimed the headlines of an article in the *New York Times* just a few years after the first satellite went up, is a prisoner in the solar system. The idea of a closed system, even if there is stress on a dynamic equilibrium within it, may convey a sense of limits which are imposed, or will or should be imposed, not by the condition of the planet itself, but by men upon other men. Thus the use of the term "dynamic equilibrium," which today we hear frequently associated with the term "steady state" (where "dynamic" is used to counteract the unwelcome connotations of the word "steady"), may nevertheless lose all power to convey any idea of movement or change.

When the Society for Applied Anthropology adopted its Code of Ethics in 1952,[1] the words "dynamic equilibrium" conveyed the idea of a good and desirable state, but by 1962 younger members were rebelling against the term as a re-

* The Society for General Systems Research, of which Dr. Mead is past president. [Ed.]

21

actionary idea. The rebellion was the sharper because those who rebelled were not fully conscious that they were living in a period when the possibilities of space exploration and space travel were seen as providing man with almost infinite possibilities.

In any period of history and within any culture, the way in which the use of models is phrased, and the contexts in which they are advocated, must be scrutinized very carefully. An ecosystem may suggest a set of existing rich opportunities which have only to be conserved, cherished, and developed, or a set of limitations so inexorable that the slightest transgressions will upset the system, irrevocably, as an island is overrun and destroyed by rats or rabbits, fertile soil soured forever by the indiscriminate use of water, the aquifers in which water has been stored for aeons, tapped and exhausted, and the power of a given environment to produce a life-support system reduced in ways that are disastrous to man's well-being. It is only a step from this latter position to conceive of acts of destruction of the ecosystem as sin, as well as crime, and to find the poetic phrasing of nonagricultural American Indians when they first encountered the plow as sacrilege, proclaiming that the plow had ripped open "our mother's breast" and the scythe had cut off her tresses.

The title *Limits to Growth* has presented such a dilemma to a world audience.[2] To the young, the dispossessed, and the poor and ambitious of the world, it has an ominous sound, a sound in no way diminished by the recommendations which are drawn from it. Limits to growth become limits to *my* growth or to *our* growth, limits imposed by *them* upon *us*, a plot of the establishment, of the rich countries, the haves against the have-nots; the sacrifice of those who have not, to a Moloch of the present high standard of living—in the few countries that attained such a standard before the unlimited use of nonrenewable resources was seen as courting disaster for the whole world. Growth becomes an easy symbol for well-being; the growth of children and trees is invoked in the image, which has been used by expansionist economists to link together economic exploitation of the earth's resources and the construction of more and more power plants, with national growth and well-being. Yet it is also only a small step of the human imagination to shift growth, the condition desired for all living things, into a monster, in which a tumor is referred to as "a growth," unwanted in any case, and in danger of becoming malignant. So in the polemics that accompany the discussion of the Club of Rome studies, while the radical young in the developed countries attack the study as setting limits on good growth, designed to bring poor countries the basic necessities of life, the representatives and advocates of the less industrialized countries slip easily from polemics about limitations on their own futures, to accusations against the monstrous and predatory economies of the industrialized countries which have devoured their resources and enslaved their peoples.

Meanwhile only too frequently the specialists in systems building brand as irrational the response of those who will not admit the validity of the model they have built, and even of any kind of model building. It is essential that we should begin to take into account the other aspects of the human mind, the poetic myth-making metaphorical part of the mind, which operates both in the systems builders and in those who are exposed to the models which they build.

Two vivid examples of such metaphorical usage, developed by natural scientists, are "Spaceship Earth" and "The Closing Circle," the latter the title of the recent book on the environmental crisis by Barry Commoner. "Spaceship Earth" was designed to convey the idea of the whole of mankind living within a closed system which is exceedingly vulnerable, and it contained the idea once expressed by the metaphor, "we are all in the same boat," and the invocation of space as the milieu of this planet and of a spaceship. But it also expresses the deep preoccupation of modern scientific man with what he himself can create. A spaceship is designed by human beings, built by human beings, and navigated by human beings—which is exactly what the planet is not. So under the guise of a pro-environmental caution, the omnipotent feelings of the engineer are thinly disguised and the whole planet, the nature of which was determined aeons before the appearance of man, is represented as man-made, and subject to whatever alterations a properly alerted engineer can introduce.

"The Closing Circle"—as a metaphor—invokes a sense of apprehension and fear, of something inexorably closing in on us, but Barry Commoner in his conclusions maintains that he is using it as a symbol of hope, as a conviction that all it will take is a little diffusion of knowledge to make human beings agree to take the proper precautions and close the circle by recycling materials and eschewing the use of those materials which cannot be recycled or which overstress the atmosphere.

World models which rely upon the inclusions of the environmental crisis all suffer from this type of ambivalent inclusion of unacknowledged and unexamined elements.

A second hazard which accompanies the use of models, in which they are used retrospectively to test their usefulness and to include the possible consequence of alternate solutions which might have been used, is that this is a type of experimental thinking that has become associated with war games; that is, the method becomes so strongly associated with the use to which it is put, that the method itself becomes discredited, and those who should be most concerned with models which may help to ensure the survival of mankind—that is, those who wish to eliminate war because they value man—instead repudiate models as a kind of war-making.

A third hazard comes from the interpretation placed upon

a model which is exploratory only—as the Limits to Growth model is—when those who have developed it proceed to draw conclusions from it which are both unjustified by the model and antithetical to the avowed aims for which the model was developed in the first place. When Limits to Growth does not explicitly allow for changes in values as a result of the model itself, the first groundwork for its rejection is laid—"freeze the world at a 1975 level" becomes salient over pious remarks about meeting basic standards for all. When Forrester suggests to the World Council of Churches* that Christians should abandon any efforts to save the dying, the starving, the victims of earthquakes, and concentrate on long-term effects, this alienates all those who know that any Christian willingness to protect the future must be founded upon Christian precepts to relieve suffering now, and they join those who attack the M.I.T. model, the Club of Rome, and the use of all models, as intrinsically destructive of human values.

All these effects are, in a sense, external to the model itself, and a function of the way in which the model builders name, use, and interpret their simulations.

When we turn to the model itself we find more serious problems. The kind of model represented by the Forrester and Meadows Limits to Growth model treats the planet—as Laszlo points out in his paper—as if it were homogeneous in resources, in ability to utilize capital, etc. This has numerous untoward consequences. All those who seek to apply the model to the situation of their own country are able to repudiate any general statement as nonapplicable, and furthermore it defeats one of the most important functions of a model, that of revealing ignorance. When data on world resources is homogenized it obscures the areas of crucial ignorance (most particularly of China). Where ignorance and the significance

---

* Cf. the chapters by Falk and Burhoe, below. [Ed.]

of ignorance should be revealed by a model of structured differential information, it is instead obscured. When the model builder contends that it is possible to calculate the amount of copper in the planet, and the model draws on such world-supply figures, then it makes it possible for the economist to escape from any difficult problem posed by the model by discussing the effect of scarcity on exploration and the lack of knowledge about degrees of access. By the time a couple of hours have been devoted to such naïvetés of the model, the economist, who should have been attending to the very points that the model claimed to make, has been reconfirmed in his distrust of the natural scientist's constructs about the consequences of seriously depleted resources.

We then come to the questions raised in criticism of the Forrester model in Professor Laszlo's paper, i.e., the lack of inclusion of other variables, the "more subtle psychological, social, political, and cultural factors."

Here we come again to a basic point of the use of models as metacommunication. The exclusion of these variables would not in itself be undesirable, if they were properly included in the *presentation* and *interpretation* of the model. If the limitation inherent in the use of the most measurable factors—and the assumption of all models must be that the factors used can in some way be coded at least into a pattern relationship—the discussion of applications can include a sophisticated inclusion of these other factors. In his concluding discussion of the Forrester model, Laszlo includes an exposition of such factors which, although couched as a criticism of the model, could equally well be handled as a way in which further discussion can be stimulated by such models.

It may well be that a refusal to use a model in any way as value-loaded or predictive may be the most desirable way in which world models, so essential to man if he is to intervene effectively in the present chaotic condition of the world, can be made and used. If models stick to hard data they can then be consciously used for a discussion of different paths of action, of whether the present disastrous handling of resources is to be corrected by a drastic reduction in the population, a "lowering of the standard of living" through the use of less energy and less hardware, a "raising of the quality of life" through the deflecting of human effort to activities which neither pollute nor exhaust nor stress the environment, through an increase in social justice in a distribution of material resources over the whole planet, or by the affluent countries arming to protect their affluence. The model can make the alternatives clear without including the values which would underlie the choice of one rather than another alternative.

But if the discussion of the implications of such a model is to take place outside and around the model, we must have more rigorous standards of the way social, political, psychological, and cultural factors are to be invoked. The capacity of a system to correct itself—discussed by Dr. Laszlo—should include two provisions. (1) We have nothing in the history of human societies that suggests that any system can be depended upon to correct itself rather than perish, and (2) the societal systems of which we are speaking are systems in which conscious purpose has been partially injected into the system in such a way that the kind of self-correction invoked by Laszlo, on a natural-systems analogy, cannot be depended upon.[3] Once a linear purposeful action of any sort—control of currency, construction of pipelines, manipulation of voters' beliefs—has been introduced into a societal system, the kind of self-correction envisioned by Dr. Laszlo is not dependable. The argument shifts to two alternatives, a return to a system which admits of no conscious purposeful intervention by man—which is impossible to envision as human beings have been purposely intervening in the ecosystem since they were able to use fire—or the progression from a partly purposeful system to a more

27

completely purposeful system. The prospect of a more complete and conscious intervention by man in his own evolution, in turn, restresses the need for simulations of activities too danger-ous, long-term, and massive for experimentation, and too com-plex for the exercise of intuition . . . even the kind of intuition to which the findings of the M.I.T. study are said to be counter-intuitive!

So in a critical discussion or evaluation of a model there is again a metacommunication, based on the assumptions made —thus Dr. Laszlo's discussion, which assumed that there are scientific reasons for believing that human beings will behave in a specified and desired way in response to an emerging, but not yet perceived, world system. This is in effect a communica-tion of optimism, which in turn may lead to the acceptance of man's interpretation because it suggests a faith in man's capacity to survive which will require no specific efforts, since man in general is able to make a "spontaneous reorientation of value-guided action patterns." Thus the implicit pessimism of the M.I.T. model is transformed into implicit optimism, which is the corollary of the principle, "conscious planification of the states of a dynamic open system is not needed," and which would then eliminate the need for models, unless models are seen as an expression of such spontaneous reorientation.

It is to levels like these that I think we should address our-selves, as well as to the technicalities of constructing world models.

3

Some Political Implications of the Forrester World System Model

ALASTAIR M. TAYLOR

"Depend upon it, sir, when a man knows he is to be hanged in a fortnight, it concentrates his mind wonderfully."

Samuel Johnson

1. Assessment of the Model

As we have seen in Professor Laszlo's paper, in its original form the Forrester model comprises five variables—population, nonreplenishable natural resources, capital investment, fraction of capital devoted to agriculture, and pollution. Each of these system elements (or "levels") represents a process of accumulation, while "flow rates" cause the levels to change. All levels are interlocked by feedback loops and the consequences can be plotted in a time sequence. Given the growth behavior of this world model's five system levels, the megacomputer was unable to come up with any combination of conventional responses that could provide enduring solutions. In short, on the basis of the work of Forrester or, again, of Dennis Meadows and, independently, of scientists and technologists in the United Kingdom, human survival in the next century appears to require a drastic shift from exponential growth to steady state during the decades ahead. The earth is finite.

I agree with Professor Laszlo that the Forrester model represents a useful introductory, but grossly simplified, model of the global ecosystem, so that any of its scenarios must be evaluated with caution. Among the criticisms directed against it has been the model's use of too few variables (though later versions have incorporated additional elements, but without effecting significant changes in the original conclusions). I could envisage increasing the variables to include energy production and consumption, communication flows, transportation networks, levels of literacy and technical education, national and regional rates of urbanization, and societal structures according to occupation, class, administrative organiza-

31

tion, and type of government. Again, the Forrester model lumps all natural resources together without considering resource substitution or the effect upon the global system as one or more becomes depleted, nor does it consider new forms of energy (fusion, hydro, solar, and wind power) which would appear to be of infinite duration and nonpolluting. There is also criticism that much more knowledge and study is required of the coefficients or "multipliers" which give the effect of each level on the others, and the extent to which coefficient variations will affect the model's predictions.

In my view, a major deficiency of the present model stems from its "black box" approach: no recognition is made of regional variations or of the need to distinguish between the respective living standards and economic and technological requirements of the developed and underdeveloped countries. The computer's scenario calls for an across-the-board reduction in population, investment, and resource usage which, if implemented, would create a steady state world system in which present disparities in living standards are essentially maintained. Such a scenario must remain completely unacceptable to the peoples of the Third World who would regard it as a sophisticated and permanent application of "neocolonialism."* If steady state in any form is to be realized, we shall have to accept continued growth in the Third World for quite some time ahead—which would imply that if planetary stabilization is to be achieved, a proportionately greater cutback will have to be accepted by the advantaged few, such as the United States and Canada, Europe, and the Soviet Union.

Like Laszlo, too, I question Forrester's seemingly sole reliance upon coercion to bring about the transition to steady state: by socially enforced pressures at present, or system-generated pressures subsequently. While also sharing Dr.

* Cf. Mead, above. [Ed.]

Mead's view that no sociocultural system is automatically self-regulating and self-stabilizing, I concur in Laszlo's recognition of the capability of social, psychological, cultural, and political factors to help transform existing behavioral patterns in accordance with the objective needs of the planetary ecosystem, and in his faulting of Forrester's "single master assumption" that "the internal forces of human systems are inescapably expansionist." True, the overall situation in modern times does appear to corroborate such a thesis. Nevertheless, we should also recognize that many societies, past and present, have never made a fetish of material growth, while planetary history is replete with examples of still others which shifted from growth to steady state and attained a viable and socially acceptable equilibrium.

Despite these criticisms, I ascribe significant value to the construction of world system models, to which Forrester has now made a noteworthy initial contribution. Concerning the argument that its projections are based on incomplete evidence so as to invalidate the conclusions, it is important to distinguish between "predictions" in the sense of prophecy, and "forecasting" as that term is employed by, say, meteorologists, who undertake to calculate future events on the basis of a rational analysis of existing data. For that matter, as I have stated elsewhere, "science has never had all the relevant data, so that replacement of one conceptual model by another has occurred as further evidence became available. Yet even Ptolemy's geocentric model of celestial mechanics proved a useful initial step in the construction of the Copernican heliocentric model, which was in turn refined by the inputs of Kepler, Galileo, and Newton. Moreover, governments are daily having to make decisions on insufficient data because they cannot afford to wait until all the evidence is in—and who is to decide what is sufficient?"[1] In the final analysis, all criticisms of the projections made by Forrester, or again by Meadows and others, have to be measured against an invariant in the man-environment nexus, namely, growth cannot continue indefinitely on a planet which in size and resources is finite. To this proposition I would add a second: a small proportion of the world's population is consuming a massively disproportionate share of these resources, while contributing a similar share of ecological pollution—and it is unrealistic to assume that the remainder of the world's peoples will continue to accept indefinitely their present ratio of resource usage.

The construction of world dynamic models can serve a further purpose at this juncture. Apropos of our opening quotation from Dr. Johnson, the prospect of ecological disaster in the decades ahead may hopefully "concentrate our minds wonderfully" so as to set remedial programs in motion. In Johnson's day, the doomed man might still have obtained a last-hour reprieve, and we may not have to pay the supreme penalty either. But to gain such a reprieve without any collective effort on our part assumes either that the model's scenarios are so wildly in error as to nullify much or all of what has been said about exponential growth, or that science can be relied upon to bail us out of any global predicament—such as by replenishing exhausted stocks of presently nonreplenishable natural resources. Personally, I prefer prudence to either simple faith or more sophisticated forms of inaction.

2. Implications for the Nation-State Paradigm

The equation of growth with economic prosperity, political power, and social well-being is part of a Western tradition that is closely associated in turn with the nation-state system. The Eotechnical stage had devised a technology for revolutionizing medieval agricultural methods, thereby increasing the food supply and, subsequently, population, as well as helping to stimulate commerce and revive urban life in the West. Additional forms of technology—including gunpowder, printing, and maritime technics (such as the compass and other navigational innovations)—initiated the Oceanic Era. As a consequence, the burgeoning Atlantic seaboard nation-states were able to secure economic and political control of resources and populations on a global scale, flooding Western Europe with a veritable cornucopia of overseas territories, precious metals, raw materials, hitherto unknown foodstuffs, new markets, and additional reservoirs of manpower (including slaves). The normative implications of the Forrester prognosis run counter to some five centuries of Western growth and expectations.[2]

Specifically, the Forrester conclusions challenge the fundamental tenets of the societal ideology which has dominated that nation-state system since the end of the eighteenth century. One might cite the respective experiences of the American and Canadian peoples in filling out their continental spaces. In doing so, despite their historical and environmental differences, they shared an ideological paradigm that included such institutions as representative government, popular education, equality before law, and economic liberalism, i.e., the free enterprise system. Geographical expansionism and economic liberalism conjoined in the popular assumption that growth is to be correlated with progress, measurable in such terms as railway and road mileage, capital investment in industry, acreage under cultivation, population increase, and gross national product—in short, that the quantity and quality of life are mutually supportive, if indeed not actually synonymous. The history of this process of growth demonstrated a character both exponential and seemingly open-ended. As a consequence, the assumption of never-ending growth has become deeply immersed in the psyches of American and Canadian societies, an assumption unquestioned in turn by political parties in either country.

The historical role of the nation-state paradigm in the

acquisition and usage of natural and human resources must be understood before we can suggest a different model that could assist in transforming societal structures and values, thereby creating a new equilibrium with our finite, but infinitely precious, planet.

a. *Genesis: the State as Primary Actor*

The state as primary actor is marked by (i) the claim to "sovereignty," that is, the acceptance of no secular authority superior to itself, and (ii) the claim to unfettered proprietorship and control of delimited territory and the population living therein. The nation-state system matured within a model of Cartesian spatial coordinates and Newtonian mechanics. Occupying a specific two-dimensional plot of the earth's surface, a state was held to be independent of all other states and absolute in its juridical pretensions. In conceptualizing the norms of international behavior, jurists of the period developed their concepts of war and peace, aggression and defense, upon a "spatialized sovereignty" based upon one-to-one—that is, isomorphic—relational thinking.[3] Thus, each state (as a person in international law) possesses certain essential characteristics: a defined territory, a permanent population, a government, and a capacity to enter into relations with similar polities.[4] All states possessing these attributes qualify for "sovereign equality" (to employ the terminology of the United Nations Charter, Article 2[1]). As a consequence of this juridical isomorphism, positivist law accords states equality of status while ignoring all relativity of stature, which comprises such variables as differences in population, resources, size and location of territory, economic organization, technological development, education, and so forth. This is the metaphysical foundation on which the modern nation-state system is founded.

36

b. *Geoeconomic Constructs*

Because a state possesses territory, people, and a government responsible for the authoritative allocation of that state's physical and human resources, we can perceive an inseparable nexus existing between environmental and societal factors in the formation and functioning of a state (or any other kind of polity). This nexus is responsible in turn for various economic and political activities and institutions which we might more specifically designate as "geoeconomic and geopolitical constructs."

In the Oceanic Era, the Western European nation-states discovered and claimed vast overseas domains which they regarded as appendages to the metropolitan sovereignty. These were contained as completely as possible within a closed system of occupancy and trade, whereby the colonies shipped raw materials to the imperial power while at the same time also serving as a controlled market for the latter's manufactured products and other goods and services (as well as for the disposal of surplus or unwanted elements of the metropolitan population). This geoeconomic system, known as mercantilism, was superseded in the eighteenth century, notably by a new economic doctrine, propagated by Adam Smith, of laissez-faire, which held that a nation's true wealth lay not in precious metals or colonial preserves, but in maximizing production without governmental interference, and also in unfettered freedom of trade among all countries. *The Wealth of Nations* appeared the same year as the Declaration of Independence, and both helped sound the death-knell of mercantilism. However, the last word on the utility of colonies had not been said. With the advent of the Industrial Revolution and the unprecedented expansion of European rule over a major portion of the globe during the last century, the accelerating demands for raw materials, markets, and areas for investing surplus capital went hand in hand with the scramble for colonies in Africa, Asia, and Oceania. The tide of imperialism has receded dramatically in our lifetime, but the transfer of sovereignty to erstwhile political dependencies has not been accompanied by a concomitant transfer of economic independence and freedom of initiative. As a result, hundreds of millions of people in the underdeveloped regions of the world have still to contend with the latest geoeconomic phenomenon, known familiarly as "neocolonialism."

c. *Geopolitical Constructs*

At this point I come to a major aspect of this paper. The political organization of planetary space has given rise to a number of theories and geopolitical models. I shall discuss them shortly, but meanwhile will suggest their inclusion of certain fundamental concepts.

(i) *Social Darwinism:* The potency of Darwinism in biology led to its rapid transposition to numerous other spheres in turn, a phenomenon known as Social Darwinism. Thus, Darwin's theories were exploited in efforts to legitimize such varying doctrines as laissez-faire capitalism, Marx's class struggle, nationalism, militarism, racism, and eugenics. For his part, Oswald Spengler contended that no treatment of history "is entirely free from the methods of Darwinism—that is, of systematic natural sciences based on causality." To him all cultures were organisms, so that the "great problem set for the twentieth century to solve" was "to explore carefully the inner structure of the organic units through and in which world-history fulfills itself, to separate the morphologically necessary from the accidental and, by seizing the purport of events, to ascertain the languages in which they speak." Spengler maintained that "every culture passes through the age-phases of the individual man."[5]

Social Darwinism played no less a central role in German

geographical theory. Prior to Spengler, two geographers had also advanced organismic theories. Karl Ritter was an environmentalist who conceived of our planet as a living globe, with the continents comprising the primary organs of that planetary organism. For Ritter, "the unity of the earth, the unity of the continents, the unity of every physical feature of the continents, and the building up together in a perfect symmetry and mutual adaptation of parts, is the crowning thought of Geographical Science."[6] Pursuing Ritter's concept further, Friedrich Ratzel developed the theory of an "organic state"—a living entity existing in space. *"Jede Staat ist ein Stück Boden und Menschenheit"*[7]—and applying the Darwinian hypothesis of the survival of the fittest, Ratzel argued that the state is involved in an endless struggle for space, or "living room" (*Lebensraum*). Because of the necessity of growth for any organism, it must take place, if need be, by force. The expansion or, conversely, contraction of political space attests to the health or decay of a state.

The Swedish geographer Rudolf Kjellén carried Ratzel's thought to the point where he envisaged the state as both a living organism and equipped with moral and intellectual capacities. In this century there emerged in Germany the doctrine of *Geopolitik*—applied geopolitics—by which the concept of the organic state was linked to military objectives and logistical strategies—and of course "justified" by appeal to Darwin's natural selection and the preservation of favored races in the struggle for existence. Under Haushofer and his group (at the Institute for Geopolitics at Munich), *Geopolitik* served as a weapon in Hitler's arsenal to further the cult of Aryan racism and extend German territorial suzerainty.

That a scientific theory, Darwinism, can be employed to support doctrines that are either contradictory, intellectually questionable, or morally repugnant to many who accept the functioning of "natural selection" within Darwin's own frame

39

of reference, raises a crucial question. Does a fallacy inhere in transposing to human societies concepts which were originally formulated to explain phenomena restricted to biological organization? As a corollary, are human actions *sui generis*, i.e., indeterminate to the point where they are impossible to order and analyze under any overarching concept capable of integrating the various orders of nature? Our answer to the first question is affirmative, to the second question negative— and in both instances that answer is grounded in general systems theory. The principle of isomorphism enables physical, biological, and sociocultural phenomena to be compared systemically, while the application of the principle of integrative levels demonstrates where the fallacy of Social Darwinism lies—in reductionism. Since each level of organization possesses its own characteristic structure and emergent properties, to reduce a more complex to a simpler level involves losing what was unique to that higher stage. Even as the organization of organisms is more complex and heterogeneous than non-biological systems, so in turn human societies make use of the physical and organic levels while, in addition, functioning at new and still more heterogeneous levels of organization and integration. Hence, to categorize a human society as an "organism" is to lose those very qualities of structure, self-regulation, and self-direction unique to the sociocultural plane, including all the conscious teleological and axiological factors which are exclusively its possessions. Yet it is precisely this exclusion of ethical and moral elements which enables the Social Darwinists to advocate the rule of force, war, and territorial expansion in the name of survival of the fittest.[8]

(ii) *Power Paradigms:* The use of biological arguments to justify a nation-state's growth in competition with other political "organisms" is compatible with the advocacy of power to attain national strength and prestige and, if and where necessary, to engage in territorial expansion. We may define "power" as the capacity to act; more specifically, as the capacity to control other elements within the given political system or its environment. As a consequence, the application of power in the form of physical force is regarded as the *ultima ratio* in both intra- and international relations. Power politics has had many advocates in various cultures: Kautilya in ancient India; Machiavelli in Renaissance Italy; and Hobbes in Stuart England. In contemporary America, we find a preoccupation with power as the most important and compelling element in state behavior among political scientists such as Hans Morgenthau, who delimits it as "man's control over the minds and actions of other men," while "in international politics in particular, armed strength as a threat or a potentiality is the most important material factor making for the political power of a nation."[9] Henry Kissinger, who has played a key role as power broker for President Nixon, insists that "no idea could be more dangerous" than to "assume that peace is the normal pattern of relations among states. . . . A power can survive only if it is willing to fight for its interpretations of justice and its conception of vital interests."[10] The graduated use of power, extending to the thermonuclear level, has been the concern of such "think tanks" as the Rand Corporation and the Hudson Institute (to say nothing of the occupants of the Pentagon and the spokesmen for the aerospace and electronic warfare "subcultures" of the military-industrial complex in the United States).

These power paradigms place strong emphasis upon the interplay of political, logistical, and technological factors. It is very often the technological variable which is found to be dominant in such models. For example, A.T. Mahan in 1890 published a persuasive treatise on the influence of sea power in history, in which he argued that maritime technology had decisively swayed the destinies of nation-states throughout the period between 1660 and 1783. During the nineteenth century,

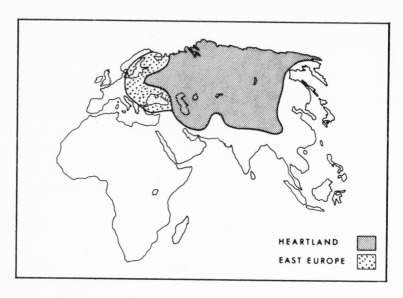

Figure 1. Mackinder's Heartland Thesis (1919)

however, the application of steam to land transportation resulted in the railroad era, which in turn opened up the hinter-
lands of all the continents. Recognition of the geopolitical
implications of this technological advance came in Sir Halford
Mackinder's Heartland thesis, which appeared in its original
form in 1904.[11] This geophysical theory held that whereas in
the previous generation, "steam and the Suez Canal appeared
to have increased the mobility of sea power relatively to land
power," transcontinental railways had now created in the vast
"closed heartland of Euro-Asia," a pivot region inaccessible
to navies. Outside this pivot area lay a great inner crescent,
comprising Germany, Austria, Turkey, India, and China, and
an outer crescent, composed of Great Britain, South Africa,
Australia, the United States, Canada, and Japan. Mackinder's
thesis argued that the central core of the Eurasian land mass
was impregnable to attacks launched by maritime powers, so

42

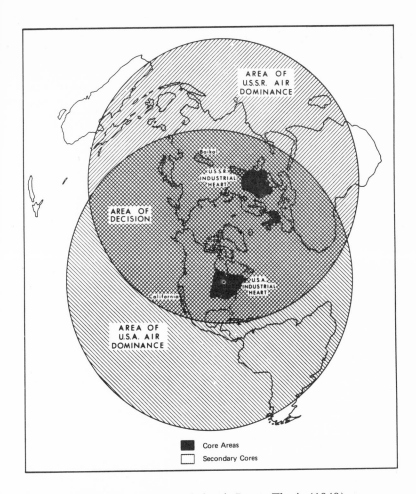

Figure 2. De Seversky's Air Power Thesis (1949)

that whoever held it possessed the key to global supremacy.[12]

But even as technological advances had been utilized by Mackinder to supersede Mahan's sea-power thesis, so the advent of aviation with its addition of a third dimension to man's environmental control capability served to outmode the Heartland concept in turn. Employing a polar map projection,

43

Alexander de Seversky argued that the United States and the Soviet Union dominated immense areas within their respective orbits, with an overlap existing between them in the polar region. He called this overlap the area of decision, and contended that its air control could ensure global strategic dominance.[13] But yet another technological increment was to destroy this new geopolitical thesis as well: intercontinental ballistic missiles and Polaris submarines have left no portion of the planet's surface immune to direct atomic attack.

(iii) *Mechanical Equilibrium Paradigms:* Various theories emphasizing power as the central element in societal relations also employ simple mechanical equilibrium models, familiar to nineteenth century physics and engineering. This is scarcely surprising. The Newtonian world view perceived the universe as a great machine, with all the processes of nature and man alike reducible to matter and motion and accountable in mechanical terms.[14] Consequently, we still encounter the "balance of power" thesis, elaborated at length in Hans Morgenthau's *Politics Among Nations* (with its apposite subtitle: *The Struggle for Power and Peace*).

Whereas the individual nation-state seeks to maximize its own power—often by territorial aggrandizement—this tendency stimulates in the larger political environment countervailing action to maintain or restore an overall equilibrium. Hence the balance of power whereby concatenations of atomistic national policies engage in alliances so as to aggregate their individual strengths in the name of self-preservation. In the eighteenth century, Vattel could write that Europe had become "a sort of Republic" whose members arranged their affairs "so that no state shall be in a position to have absolute mastery and dominate over the others."[15] As Quincy Wright points out, the balance of power has been traditionally preserved by the following methods: ad hoc alliances; permanent guarantees to a particular state in a strategic position

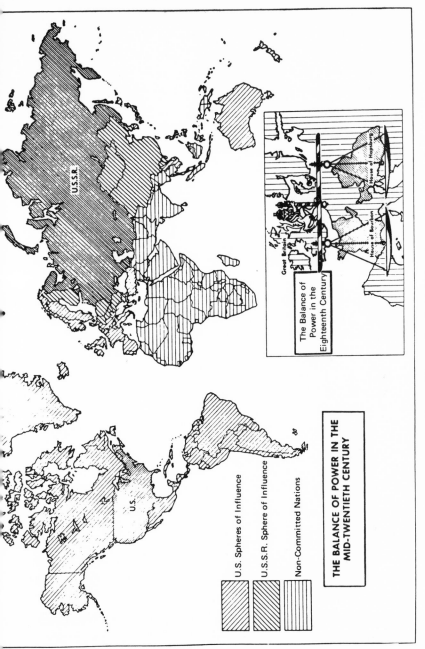

Figure 3. From Morgenthau's *Politics Among Nations*

(often as a buffer between two powerful states); regional arrangements (collective security within limited areas, such as the Monroe Doctrine); and more universal systems of collective security (such as the League of Nations or the United Nations). He adds that the concepts of balance of power and collective security have different fundamental assumptions. "The balance of power depends upon national policies of ganging up against the overpowerful, whereas collective security depends on an international obligation to collaborate against any aggressor. A government cannot at the same time behave according to the Machiavellian assumptions of the balance of power and the Wilsonian assumptions of international organization."[16] However, none of the collective security organizations has yet been able to subordinate the balance of power to its juridical and ideological postulates. "Consequently, they were not able to survive serious disturbances of the balance of power."[17]

The mechanical equilibrium model, pervasively employed to explain the behavior of power in the nation-state system which had matured within a Newtonian world view, was to make its appearance in turn in geopolitical theory as it shifted from its traditional descriptive and morphological approach to one more functionally oriented. This emphasis upon process was developed by Richard Hartshorne, who viewed the state as seeking "to create a *region* of high degree of functional unity and in certain respects of high degree of uniformity."[18] Since every state possesses regional differences and similarities—differences that tend to disrupt and similarities that tend to bind together—its viability must depend upon the relationship of "centrifugal" to "centripetal" forces. Among centrifugal factors, Hartshorne included physical and human barriers as well as intra- and interstate regionalism, while "the basic centripetal force must be some concept or idea justifying the existence of this particular state incorporating these particular

46

regions; the state must have a *raison d'être*—reason for existing."[19] Here Hartshorne drew upon the concept of the state-idea as developed by Ratzel, who had contended that those states are strongest "in which the political idea of the state fills the entire body of the state, extends to all its parts."[20] (Jean Gottmann subsequently extended this concept of state-idea to that of iconography, "the whole system of symbols in which a people believes."[21])

Hartshorne applies his functional thesis also to a state-area's external relations. Again he perceives the interplay of integrating and disintegrating forces as they function in the state-area's territorial, economic, political, and strategic relations with other areas. The theory consistently employs a simple push-pull thesis. We might note that all the geopolitical models thus far discussed consider the nation-state as the primary actor, and hence the international political environment as comprising nation-states (together with their territorial dependencies). This paradigmatic approach to the sociopolitical ecumene is essentially "Ptolemaic," that is, each state regards itself as the central body in geopolitical space, with other bodies rotating around it (preferably as satellites). Within this paradigm, power as employed by the nation-state—either independently or in concert with other states—remains the self-justifying arbiter in the geopolitical environment.

But how valid is this paradigm in the light of technology's sequential—and accelerative—supersession of Mahan's, Mackinder's, and de Seversky's theories? Surely the Space Age has rendered all but obsolete not only the two-dimensioned nation-state's traditional strategy but, concomitantly, its paradigmatic pretensions to independence and sovereignty as well. Once more we find ourselves nearing the end of a long-standing mode of thinking, and this new conceptual awareness is in consonance with the Forrester conclusions in the ecological sphere. Since the earth is finite and every point on its surface can now

47

be programmed for thermonuclear destruction, exponential growth in military technology has also to be brought to a halt. Such growth has already achieved "overkill" and a state of affairs that is financially ruinous, ecologically poisonous (given the fallout of atomic bomb testing), and logistically counter-productive. The nature of power itself has been radically transformed in the sense that its thermonuclear potency dare not be accepted as a viable option in the decision-making process—so that the United States, for example, found itself a muscle-bound giant held to an inconclusive (and for its superpatriots a humiliating) outcome in Vietnam by an infinitesimal portion of the world's population and resources. This dual transformation of power and the nation-state (as traditionally conceptualized) reinforces the logic of arms deescalation, and the further move to steady state through the creation of new conflict-control modalities, such as international peace-keeping.[22]

3. Towards a New Geopolitical Paradigm

Ours has been a century distinctive for new concepts regarding the fundamental nature of space, for technological innovations in the physical use of space, and also for the advent of "closed space" in the political sense. Thus, Einstein conceived of a new model of celestial mechanics calling for a time-space manifold that discarded Newton's concept of space and time as limitless and separate, and which employed Riemannian geometry to account for phenomena in "curved" space. Meanwhile, the Wright brothers and other pioneers provided mankind with a three-dimensional environmental control capability. This technological quantum created travel and communications *continua*. For the first time, movement was no longer impeded by geomorphological discontinuities on the earth's surface (in the form of rivers, seas, mountain ranges, deserts, and polar ice masses), while the transmission of an electrical signal girdling the equator faster than seven times per second had laid the foundations for McLuhan's "global village."

Political space has had to be conceptualized anew, i.e., in three dimensions—more specifically, in terms of both "inner" (or subjacent) space, and "outer" space, which theoretically extends to the most distant galaxies. Of crucial concern at this juncture in mankind's planetary evolution is the following question: What happens to geopolitical "property" and traditional pretensions to unlimited and unfettered freedom of unilateral action based upon a two-dimensioned concept of spatialized sovereignty? This question is made even more relevant by the filling-up of political space everywhere on the globe's surface. After Peary and Amundsen reached the two poles, history's hitherto unbroken era of *terra incognita* had come to an end. By this time, furthermore, the last of the "free land" in the continental United States had been taken up. And although the Statue of Liberty continued to welcome Europe's "poor and huddled masses yearning to breathe free," the screening of those masses began on nearby Ellis Island, while, following World War I, laws were enacted that, on a quota basis, drastically restricted immigration. With these initial shifts from open-ended growth towards restriction of alien population, American society had begun to recognize that its own continental confines were both territorially and demographically finite—and this sense of finitude was in turn underscored by a rapid and prodigal depletion of once seemingly inexhaustible natural resources. In short, the stage of continental steady state has already been reached in the American experience.

In recent years, much innovative work has been done in such areas as territorial behavior and spatial apperception. We are all aware of the proliferating literature from ethologists on the features and functions of animal territoriality. This work elicits two closely associated questions. First, how far can we assume comparability between animal and human territorial behavior? Secondly, is territoriality in man instinctual and essentially ineradicable except through major evolutionary change, or is it culturally derived and therefore susceptible to alteration through socialization and culture change? As Soja and others point out, these questions are highly relevant to the nature of aggression in the human species, and also to man's psychological relationship with his ecology (and hence whether his tendencies—innate or acquired—impel him to regard that relationship in terms of "conflict" and "control" or, alternatively, of symbiosis and synergy).[23]

The nexus between spatial apperceptions and those societal paradigms which symbolize, and in turn largely shape, our constructs of reality, has been tardily appreciated in political and geographical literature alike. Soja has pointed out that Western perspectives on the political ordering of space have been "powerfully shaped by the concept of *property*, in which pieces of territory are viewed as 'commodities' capable of being bought, sold, or exchanged at the market place."[24] I have suggested that whereas the national state system matured within a model of Cartesian spatial coordinates and Newtonian mechanics, the changing conceptions of space in physics away from the absolute or "container" theory have profound implications for the conceptualization of political space.[25] Earlier, the Sprouts stressed the importance of environmental apperception in the foreign-policy-forming process: "What matters is how the policymaker imagines the milieu to be, not how it actually is."[26]

In an endeavor to link political processes with spatial perspectives, Stephen Jones developed his "Unified Field Theory."[27] As he points out, the term "unified field" is not based on any analogy from physics but seeks rather to unify the concepts of Whittlesey, Hartshorne, and Gottmann, and unite them in turn with political-science theory, notably Karl Deutsch's earlier studies of growth and integration among political communities.[28] Jones made use especially of Hartshorne's concept of state-idea; he perceived "idea" and "state" as two ends of a chain, connected by three other links, one being Gottmann's "circulation," which he now termed "movement." The chain attempts to account for the progression of a political idea through a sequence of actions and interactions to its implementation in the form of a politically organized area. This five-stage progression can be diagrammed and exemplified as follows:

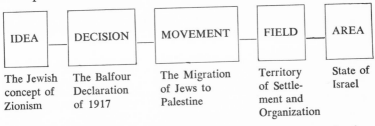

IDEA	DECISION	MOVEMENT	FIELD	AREA
The Jewish concept of Zionism	The Balfour Declaration of 1917	The Migration of Jews to Palestine	Territory of Settlement and Organization	State of Israel

Jones visualizes this process "as a chain of lakes or basins, not an iron chain of separate links. The basins interconnect at one level, so that whatever enters one will spread to all others." Moreover, the field theory can apply both to organized and unorganized areas ("like the Mediterranean, which is undoubtedly a political field").[29] As Jones points out, the flow from idea towards area is one of controlling or creating, whereas the reverse flow comprises "conditioning" (inasmuch as the existence of, say, a movement or field conditions what may take place in the basins lying idea-ward). However, the model's flow is fundamentally unidirectional (from idea to area) since this is the only way by which actions and interactions can serve to "control" or "create." In short, while his "field" theory pointed the way to the application of a systems approach to political geography, it "lacks . . . the feedback mechanism to be contemporary."[30] Though a seminal concept, the Jones model was not rigorously analyzed or further developed. It has the virtue of emphasizing "process" in the formation of all kinds of polities, and not solely the nation-state. However, this emphasis upon sociopolitical process lends itself to a more comprehensive development and methodology within a full-fledged systems framework.

Inasmuch as political science and international relations have traditionally focused upon the sovereign nation-state as the primary actor in both the domestic and global environments, in such a state-centric paradigm political geography has for its part concentrated upon the nation-state as organizer and delimitator of political space. However, the emergence of new transnational issues and perspectives have required fresh conceptual and methodological formulations in turn. Thus, empirical evidence abounds of an accelerating proliferation of transnational economic processes and multinational business enterprises; of transnational networks in science and other areas of intersocietal transaction; and of intergovernmental and international nongovernmental organizations continuously performing a multitude of functions, thereby affecting both the character of the global environment and our apperceptions of political structures and behavior.[31] Similarly, race relations as a transnational phenomenon have become progressively recognized and researched.[32] Again, postwar decolonization on a global scale has required investigation of political processes in non-self-governing territories, their transformation into sovereign polities, and the comparative study of developmental politics.[33] At the other end of this developmental spectrum we find a different transnational process at work: voluntary integration among a number of Western European states to form the European Economic Community, raising the possibility in turn of eventual federative political integration. This and similar "community" developments in other continents have stimulated a large literature devoted to integration theory and research. As Haas points out, "The study of regional integration is unique and discrete from all previous systematic studies of political unification because it limits itself to *noncoercive* efforts. . . . The study of regional integration is concerned with tasks, transactions, perceptions, and learning, not with sovereignty, military capability, and balances of power."[34] It calls also for the examination of such dimensions of interstate integration as intergovernmental cooperation, political amalgamation, and what has been described as "mass community."[35] In summary, economic, social, and technological factors are combining to shift our paradigmatic focus from the nation-state—as conceptualized heretofore in formal juridical terms and occupying no less formalized two-dimensioned political space-containers—to transnational processes within a global environment. This environment can be, in turn, also viewed as a supranational geopolitical system, coterminous with Forrester's ecological—or world dynamics—system.

4. A General Systems Model for the Forrester Thesis

It is hardly surprising that the increasing use of the term "political system" has been accompanied by an ever-growing systems-oriented literature in political science and international relations. Within systems analysis, some scholars have focused upon a "functional" approach, that is, they have been concerned with the needs or requisites of political systems, and the structures (institutions, activities, etc.) which fulfill them.[36] Again, general systems theory has been applied to different political systems for the purpose of locating structural and behavioral isomorphisms and to study the interaction of these systems with their societal environments by means of feedback processes.[37] (Not surprisingly, either, systems theory has engendered its quota of critics who have questioned both the inherent value of such an approach and the operational results obtained by its practitioners.[38])

The systems approach has also affected geographical theory, though Harvey pointed out (in 1969) that it had "not gone very much beyond the stage where we are exhorted to think in terms of systems."[39] Subsequently, however, systems analysis has been applied more rigorously to the study of physical processes,[40] to spatial analysis,[41] and, again, to urban problems and planning.[42] However, Harvey's stricture still largely applies to political geography. True, Soja seeks to relate behavioral variables—such as concepts of space and territoriality—to support his advocacy of a functional (spatial) rather than formal (areal) approach to the subject so as to enable political geography to become more process-oriented.[43] A subsequent model by Cohen and Rosenthal also emphasizes behavioral variables, but within the same areal perspective adopted by Jones earlier in his "idea-area chain." As with the latter, we find no concept of feedback and the same unidirectional vectoring from "societal forces" to "landscape."[44] In any case, we are unable to obtain from existing literature in political geography a systems paradigm appropriate for our present purpose to assess in geopolitical terms the ramifications of the Forrester world-dynamics thesis.

David Easton has been described as "the first political scientist to analyze politics in explicit system terms."[45] In his works on systemic structure and behavior,[46] he is concerned with the continuous interaction between a given political system and its environment. Thus, the latter's inputs take the form of *demands* and *supports*—for example, demands for allocations of goods and services, the regulation of societal behavior, participation in the political system, and communication of policy intent by the system's elites; while supports take such material form as the payment of taxes and military service, as well as obedience to laws and regulations, voting and other forms of political activity, and respect for public authority. The political system works upon these inputs, as well as those

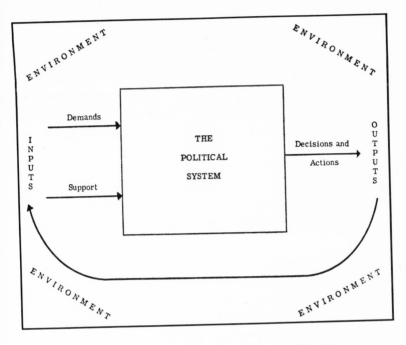

Figure 4. Easton's Simplified Model of a Political System

factors or "withinputs" generated within the system itself, and converts them into outputs, which take the form of "decisions and actions." More specifically, Easton refers outputs to "those kinds of occurrences . . . described as authoritative allocations of values or binding decisions and the actions implementing and related to them."[47]

As the diagram indicates, feedback loops account for the equilibrating processes at work; by being endowed with feedback and the capacity to respond, the political system "is able to make some effort to regulate stress by modifying or redirecting its own behavior."[48] Easton's "environment" is both intra- and extrasocietal, i.e., he differentiates between that part of the environment which lies outside the boundaries of a political system and yet within the same society, and that part which is external to the society of which the given political system itself comprises a social subsystem. (This extrasocietal environment contains international ecological, social, and political systems.) Unlike the great majority of models discussed previously, Easton's "political system" is not limited to the nation-state category; isomorphisms exist among political structures, activities, and roles whether we are dealing with tribal systems or the modern territorially oriented nation-state.

In the Easton model, as diagrammed above, feedback is not differentiated as negative or positive. Given his remarks about the need of the political system to "regulate stress," for the authorities to ensure continued inputs of support and to fulfill demands, as well as for the aim of the system to persist, we derive the impression that Easton conceptualized his model primarily in terms of *negative* feedback processes. This impression would be in keeping with the objectives of the contemporary Western political paradigm, in which the nation-state seeks to maintain overall societal stability, a pattern of evolutionary or "adiabatic" adaptation to new environmental circumstances, and in which continuous economic development—and GNP growth—is regarded as normative (so that any contrary view comprises a deviation from the prevailing societal goals and norms of behavior). On the other hand, if we want to account for quantum societal and systemic transformations—which we believe are occurring at an accelerating rate in the contemporary global environment—we have need of a model that recognizes the ways and extent to which *positive* feedback processes can become dominant.

Systems models have tended to be conceptualized "horizontally," that is, they are designed either to examine a single socio-political system or, again, transacting systems in a similar state of structural development and behavior. What we have also to account for, however, are "vertical" shifts of societal organization so as to explain increasing complexification and heterogeneity—as in today's world. Specifically, we require a systems model which, in addition to distinguishing between positive and negative feedback processes, employs the principle of integrative levels and thereby seeks to explain not only the interplay of these two types of feedback processes at any one level of sociocultural organization, but the circumstances in which quantization occurs from one level to another.

Hence, we have constructed a model which has been diagramed to account for both (a) systemic levels of sociocultural organization, and (b) cybernetic processes that demonstrate (i) systemic self-stabilization within a given level of organization and integration, and (ii) systemic transformation so as to result in a sociocultural quantum leap across an environmental frontier.

In the first diagram (figure 5), the question of delineating the number and categories of organizational levels depends upon one's particular purpose. I approached this question from the basic standpoint of the man-environment nexus because, of course, sociocultural systems are "open" and exhibit feedback stabilization with their overall environment. Regarded horizontally, each level depicts transactions occurring between the physical environmental factors, the stage of manipulative equilibration (as made possible by existing technologies and science), transportation facilities, communication networks, and the paradigmatic character and organization of the given society in which the political process ("the authoritative allocation of values") functions. It is a geopolitical model inasmuch as it correlates specific levels of societal and political organization with precise stages—or dimensions—of environmental control (as shown in the expleted- and impleted-space columns).

Viewed vertically, these progressive stages of overall environmental occupance assume a geometrical sequence: *point—*

LEVELS OF ORGANIZATION

SOCIETAL LEVEL	SYSTEM OF ENVIRONMENTAL CONTROL		PROPERTIES	EMERGENT QUALITIES				
	EXPLOITED SPACE	IMPLETED SPACE		TECHNOLOGY	SCIENCE	TRANSPORTATION	COMMUNICATIONS	GOVERNMENT
S_5	Three-dimensional (extra-terrestrial)	Megalopolis ("Ecumenopolis")	BELOW +	Electrical-nuclear energy / Automation / Cybernetics	Einsteinian relativity / Quantum mechanics / Systems theory	Supra-surface: inner space systems / Outer space explorations / Surface systems / Sub-surface vehicles	Electronic transmission (simultaneity throughout exploited space)	"Ecumenocracy" (Supra-national polities) / Multi-level transaction / Sovereignty invested in global mankind
S_4	Two-dimensional (oceans, continents)	City	BELOW +	Transformation of energy (steam) / Machine technology / Mass production	"Greek miracle" / Scientific method / Newtonian world-view	Maritime technology and navigation / Thalassic and oceanic networks / Highway networks / Railroad technology	Mechanical transmission (printing) / Alphabet	National state system / Emergence of democracy / Sovereignty of state (as primary actor)
S_3	One-dimensional (riverine societies)	Town	BELOW +	Non-biological prime movers (wind, water) / Metal tools / Continuous rotary motion (wheel) / Irrigation technics	Mathematics / Astronomy	Sailboats / Riverine transport / Wheeled vehicles / Intra- and inter-urban roads	Writing	Ancient bureaucratic empires / Theocratic polities / Sovereignty of god-kings
S_2	Particulated Universal (sedentary)	Village	BELOW +	Animal energy / Domestication of plants and animals / Polished stone tools / Spinning / Pottery	Neolithic proto-science	Animal transport / Paths, village routes / Neolithic seafaring	Ideograms	Biological-territorial nexus / Tribal level of organization and decision-making
S_1	Undifferentiated Universal (Nomadic)	Cave/tent (intraterrestrial)		Human energy / Control of fire / Stone and bone tools / Partial rotary action		Human transport / Sleds / Dug-outs, canoes	Pictograms	Biological nexus (family, hunting band, clan)

line–plane–volume as man's control capabilities increase. This sequence also demonstrates the actualization of the principle of integrative levels: each such level tends to build upon the properties and societal experiences of the level(s) below and in turn contribute its own "withinputs" and outputs—which take the forms of new technologies and new societal structures, accompanied by new paradigmatic apperceptions of the man-environment relationship. We can discern progressive developments in complexity and heterogeneity (although in any one historical situation a different, or even contrary, experience may occur).[49] In summary, figure 5 provides a time-space grid, showing both societal-environmental *stabilization* when viewed horizontally (process in planetary space) and societal-environmental *quantization* when examined vertically (process in planetary time). Mankind's overall experience has been to expand anthropogeographic space concomitantly with its accelerative contraction of the temporal sequences associated with new stages of environmental control, i.e., from S_1—the food-gathering level of socioeconomic organization, through S_2—the food-producing level, and sequentially to S_5, the level to which the Forrester thesis is directed.[50]

The relation of species other than man to their environments is determined primarily by Darwinian, genetically coded mechanisms, so that the evolutionary process can be described as *adaptive* equilibration—because, while the overall process is mutagenic and open-ended, and hence exhibits positive feedback, negative-feedback mechanisms dominate in the maintenance of individual species and their members. Conversely, organisms with sensory-cognitive circuits are at the stage of *manipulative* equilibration to the extent that they possess deviation-amplifying capabilities. Applying the principle of integrative levels, we can understand why adaptive, i.e., instinctive, homeostatic equilibration can be retained at the lower levels of an organism such as Homo, who, at the same time, functions consciously at the highest, or cognitive, level in order to adapt the external environment to fit his own constructs. (In fact, the concomitant presence of negative and positive forms of feedback free him from having to devote all his conscious efforts to satisfying basic physiological needs, as his autonomic nervous system takes care of them.)

It is by "man the toolmaker" that the equilibrating process shifts progressively from a reactively adaptive to an actively manipulative role. Hence our model recognizes the central function of technology in the development and transformation of sociocultural systems from Paleolithic times to the present; in other words, the role of technology and science, i.e., "material technics," as positive-feedback processes. Concomitantly, we should also recognize the role of societal institutions and mores to maintain continuity and persistence in any sociocultural system. In this respect, therefore, "societal technics" function as negative-feedback processes so as to ensure overall stability and societal invariance under technological and environmental transformations.

Building upon the type of systemic schema employed by Easton, we have diagramed below a model which (a) accounts for (i) biospheric and (ii) sociocultural inputs from the total environment; (b) recognizes the given sociocultural system as (i) converter and (ii) comprising numerous subsystems (including the political); and (c) relates its outputs—material and societal technics—to positive and negative forms of feedback.

The diagram indicates how material and societal technics interact and, depending upon the state of the system vis-à-vis its environment, how they combine so as to result in systemic self-stabilization or, alternatively, in systemic transformation. We shall term the first systemic process "Cybernetics I"; the second "Cybernetics II."[51] The diagram also aims to show that whereas Cybernetics I, comprising net negative-feedback

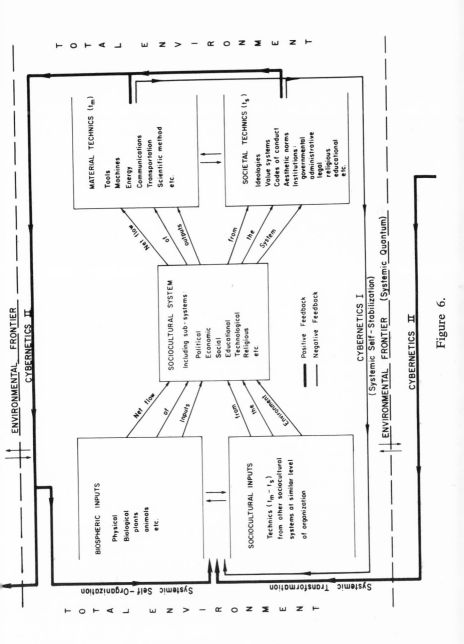

Figure 6.

61

processes, acts to stabilize a given sociocultural system within its environment, the dominant positive-feedback processes comprising Cybernetics II can (i) increase the system's negentropy and information gain, and thereby also increase its environmental control capability so as to actualize the existing potential within the system-environment nexus; and/or (ii) enable the system's outputs (in the form of material and societal technics) to cross the permeable frontiers separating one environment from another and thus quantize to a new level of societal organization. As I have stated elsewhere: "Quantization occurs when deviation is amplified to the point where no deviation-correcting mechanism can prevent the rupturing of the basic systemic framework, i.e., when the latter can no longer contain and canalize the energies and thrust which have been generated."[52]

We might briefly illustrate the application of Cybernetics I and II to actual societal experiences. Examples of Cybernetics I are subhominid societies wherein Darwinian mechanisms are fully operative; again, mature or senescent sociocultural systems in which the available material technics have achieved their maximal environmental control capability and reached steady state. Under Cybernetics II we can distinguish between two types of societal experience: systemic self-organization and development, and systemic transformation (environmental quantization). As an example of the first type (involving the actualization of the potential present in the system's technics vis-à-vis the potential and constraints present in the environment), we might choose lithic man's advancement into the high latitudes. It was made possible by the control of fire (energy production) and the invention of progressively sophisticated and efficient tools (such as microliths). This development resulted in what has been described as a "tour de force" by the Eskimos, who attained maximal use of such a technology so as to survive beyond the timberline and maintain a viable symbiosis with an austere, i.e., low-energy, physical environment. However, since environmental constraints set boundaries on expansion and control of the biosphere, negative-feedback mechanisms were to become dominant, resulting in overall societal stabilization (Cybernetics I).

Eskimo society has remained at S_1 (in figure 5). But through Cybernetics II we can see how material technics acquired in one society at a less complex level of organization permit systemic transformation to occur by means of a different environmental relationship, or in a new environment altogether. In Mesolithic times in southwestern Asia, the Natufians used stone sickles to cut certain wild grasses. The way was thus prepared for a technological quantum in this same environment, which took the form of "withinputs" in the existing societal system: the domestication of these ancestors of wheat and barley, and also of the further domestication of certain wild species of animals. The resulting "Neolithic Revolution" (Gordon Childe's term), transforming the man-environment relationship and quantizing to S_2, not only extended society's environmental control capability but created new settlement patterns, new societal organization, and a more complex division of labor, while an increased and relatively assured food supply enabled a larger population to be supported. However, at this stage of incipient agriculture, limited water supplies (as at Jericho and Jarmo) set rigid limits upon growth, so that ultimately steady state (Cybernetics I) eventuated.

We can see even more dramatically how positive feedback can bring about a quantum leap when new material technics are applied to a different environment. Whereas in Jericho and Jarmo the man-environment relationship stabilized at the Neolithic village level, the transplanting of farming technics to the rich bottom lands of the Nile, Tigris-Euphrates, Indus, and Huang-ho resulted in a hundredfold increase in food harvests, making possible a "social surplus" that unprece-dentedly increased population numbers and densities, and freed many persons to work in occupations and localities re-moved from the fields. Hence we find the rise of towns, ac-companied by more complex governmental and administrative structures, hieratic elites, etc. In short, the "Urban revolution" (S_3) describes a systemic transformation in which environ-mental control eventually covers an entire river valley and raises technological and societal activities to new negentropic levels and complexification, resulting in the first "civilizations." With this hydraulic technology, control of the river eventually reaches a new plateau of stabilization because of the con-straints inherent in the surrounding environment. Thence Cy-bernetics II yields once more to Cybernetics I, the latter reinforced by societal technics which maintain systemic equi-libration for millennia in these riverine civilizations.

Examples of the equilibrating process could in turn be drawn from S_4—and I have already discussed the geopolitical implications for S_5 from the standpoint of the Forrester thesis. Now that there is virtually no more free land (*terra incognita*), and nonreplenishable resources are being quickly depleted, the model makes clear that the traditional role of material technics must be progressively transformed from positive to negative feedback, so as to stabilize the global ecosocietal system.

5. Conclusions

It would appear that we have three options from which to choose in the remaining decades of our century:

a. *"Ptolemaic" Nationalism*

In effect, this is a continuation of the present nation-state system, in which each sovereign actor unilaterally determines its own "vital interests" and regards its relations to other nation-states in two-valued, either/or terms. ("Nations do not have eternal friends, only eternal interests.") Differences among states remain susceptible to resolution by coercion or actual physical violence—hence the probable continuation of arms races and wars. Alliances represent a concatenation of disparate, sovereign states linked in a common cause against some other concatenation. Their purpose and structure are not suited to transformation into supranational "communities." Nor do the structure, membership, and objectives of the nation-state system conduce to the mounting and implementation of the kind of global strategy required to cope sufficiently soon or adequately with the massive challenges posed by the Forrester model scenarios.

b. *"Copernican" Internationalism*

Alternatively, of course, it is conceivable that the nations of the world will agree to yield their traditional juridical pretensions and political unilateralism so as to transform the United Nations into a global government with powers not only to take executive initiatives but to enforce those decisions. Some scholars have attempted to show how the present charter could be amended so as to create such a world authority. But the organization's first quarter of a century of existence makes such a development appear highly unlikely. A very different form of Copernicanism could instead emerge. This would be the agglomeration of so much technological, military, and economic resources under the control of one or two superpowers as to make permanent their dominance—perhaps as twin stars encircled by their respective constellations of satellite dependencies. This geopolitical configuration could be equated with geoeconomic Copernicanism in the form of multinational corporations functioning throughout the world, with their headquarters, research facilities, and charters of incorporation located in a given superpower. The resulting centralization of authority to make decisions and allocate resources might well facilitate initiatives to rectify existing ecological imbalances, but this very centralization of political and economic power would almost certainly result in coercive rather than cooperative action.

c. *Multilevel Transactionality*

Based upon the principle of integrative levels, the model proposed here recognizes concurrent existence and behavior at different levels of societal organization. Such multilevel transactionality applies no less to political systems (i.e., subsystems within an encompassing sociocultural system). Already, unitary political structures—such as in Western Europe —possess two levels of government, national and municipal, while federal structures—as in North America, Australia, and India—have three tiers of governmental organization (federal, state [provincial], and municipal). Transactionality occurs both horizontally, i.e., among departments and decision-making processes in each tier, and vertically, i.e., between governmental and administrative tiers as well. Each of us lives concurrently at more than one level of sociopolitical organization. "Sovereignty" resides in all levels of the national polity simultaneously (though its ultimate invocation to legitimize action reposes with the highest, i.e., most nearly all-encompassing level of geopolitical organization). Paleolithic societies could function harmoniously with their environment by means of one level of sociopolitical organization and decision-making; the subsequent addition of new levels attests to the progressive complexification of post–Stone Age societies and their authoritative allocation of values and resources.

Viewed in this historical perspective, the evolution of geopolitical structures suggests that still other governmental levels beyond what we possess today will be required to cope with those problems in our collective environments which transcend national boundaries and are therefore no longer susceptible to unilateral resolution by individual polities. We should resist conceptualizing our options in simple dyadic terms: *either* to retain the nation-state system basically as it is *or* to scrap it and create a unilevel type of world government. Apart from the difficulties of transforming the United Nations—which was conceived primarily as a mechanism of its member states —it may well be that one or more political levels will emerge between the present national structure and any ultimate world government. Indeed, given the complexity of our global environment and its regional variations, and the large number of historically rooted national governments presently in existence, the creation of but a single supreme decision-making and enforcing authority would appear to be too centralized—and "truncated"—a step to envisage realistically. The emergence of various economic "communities" in different continents may portend the evolution of a number of regional supranational authorities throughout the world. Also, there is evidence that we may be entering into a new kind of participatory democracy—involving citizens at all levels of sociocultural organization, both governmental and nongovernmental.

Multilevel transactionality could involve more groups and individuals in the decision-making process, and also develop a new awareness of interdependence in tackling problems common to all, including the ecological challenge. For my part, I do not see traditional "liberalism"—either political or economic—moving to implement any Forrester scenario without new and appropriate conceptual inputs into our existing political subsystems, coupled with massive societal pressures. Certainly, the required transformation will have to overcome both passive resistance from the *mortua manus* of tradition, and active hostility from the entrenched bastions of power in every national society. But if this analysis has any validity, the character of those mutually supporting pillars of the nation-state system—power and sovereignty—is already undergoing a profound transformation. Heretofore, confrontation strategies have not only accepted Clausewitz's dictum that "war is nothing but a continuation of political intercourse, with a mixture of other means," but have justified the use of force as the *ultima ratio* on the grounds that in a dyadic, adversary relationship, the outcome is zero-sum, with the losses of the vanquished being matched by the victor's gains. But given our present weaponry, recourse to naked force produces non-zero-sum results, that is, all parties are made losers—whereas co-operation can bring about a non-zero-sum payoff that is advantageous to all concerned.

Finally, I would underscore a salient conclusion that derives from the application of this systems model to the Forrester thesis. The attainment, or even approximation, of planetary steady state will require a reversal of the traditional feedback roles of technology on the one hand and of societal technics on the other. Cybernetics I in time became dominant at every previous stage of societal organization and—barring a thermonuclear holocaust destroying human life itself—this process can be expected to recur in the emerging "global village" (S_5). I do not rule out a continued positive-feedback role for material technics in certain areas: growth and further environmental control will still be required on much of the planet for many years ahead, and of course we have just set our sights on extraterrestrial expansion. But henceforth a central purpose of technology must be to sustain and optimize—rather than despoil and destroy—the life-support processes of our global ecosystem. And concomitantly with this new, and highly creative, role for science and technology is required an equally new and creative positive-feedback role for our societal technics. We have scarcely begun to probe the subsurface levels of either individual or societal innovation. Such a "crossover" will call for the progressive recognition and adoption of a new sociocultural paradigm, involving different societal values and political priorities, and the mobilization of our shared resources of creativity, courage, and compassion in order to build a new world order.

4

Reforming World Order: Zones of Consciousness and Domains of Action

RICHARD A. FALK

I

Iɴ 1971 U Thant, while secretary-general of the United Nations, told the twenty-sixth session of the European Economic Commission, "If the great French philosopher Descartes were alive today he would probably revise his celebrated rules for good thinking and add two new principles: the principle of global thinking and the principle of thinking well ahead into the future."[1] Such prescriptions for relevant thought properly address themselves to matters of spatial and temporal orientation. Such explicit imperatives respond to a widespread appreciation that we live at a time of danger and opportunity, not just as Americans or Nigerians or Frenchmen, but as human beings concerned with the destiny of our species and the welfare of our planet. This aroused concern has generated an unprecedented effort to achieve a planetary perspective on human problems and prospects.

There is, as yet, no consensus about how to achieve such a planetary perspective beyond the very elementary need to detach oneself, to the extent possible, from too firm an anchorage in the particularities of time and space. One approach to universalization of thinking is to move from considerations of quality to considerations of quantity. Numbers are more objective than words, data more persuasive than opinions, charts more convincing than speculations or prophecy. The scientific temper of the times reinforces this effort to evolve a valid planetary perspective by so-called objective methods, and indeed a whole subdiscipline of sorts has grown up under the label "futurology."[2]

But somehow such exercises in quantification and projection are not nearly enough.[3] One comes away from reading Kahn and Wiener's *The Year 2000*, a characteristic effort, with a sense of having traversed an intellectual desert, which despite the methodological gamesmanship is unable to develop any orientation to the world or the future that is either useful or inspiring.[4] In the end, Kahn and Wiener are technocratic prisoners of the present, as insensitive to the future and the planet as a whole as if they had been given a severe dose of brainwashing tranquilizers by the power wielders in the advanced industrial countries who were terrified by the prospect of genuine visionary thinking. To be more specific in my indictment, the time-future of the futurologist is so uninteresting because it is devoid of either vision or empathy, it lacks both the force of human imagination or the moral strength that derives from feelings of solidarity with and sympathy for the human race as a whole.

These shortcomings also afflict, I believe, the much more significant work of the Meadows group carrying out the project on the Predicament of Mankind in association with the Club of Rome,[5] which was so successfully promoted that it engendered a mainstream backlash from those who felt threatened by the covert ideology locked in the phrase "the limits to growth."[6] Jay Forrester's public impact is almost directly attributable to the use of the computer as a basis for presenting an argument about the perils to the planet deriving from the dynamics of economic and demographic growth. Such an argument is qualitatively compelling, and in this sense is to be sharply distinguished from the inconsequentiality of Kahn-Wiener, but to connect it up with the objectivity of numbers and the authority of the computer is to engage, however unintentionally on the level of motivation, in a gigantic public-relations hoax. We do not have sufficient information to feed the computer, and we have no reason to regard its printouts as trustworthy, at least not for a long time.[7] To work toward computer credibility is an urgent task of highest priority for those of us seeking to follow U Thant's updating of Cartesian imperatives, but to hire a public-relations outfit to preach as gospel the present printouts is to participate, I fear, in the worst sorts of neoastrological games of mystification; religion is not the only opiate of the people.[8]

William Irwin Thompson has provided a brilliant critique of these technocratic forays into the future from the perspective of cultural history.[9] Thompson regards Aurelio Peccei, the founder of the Club of Rome, as an example of a "new postindustrial manager" who has the "institutional corporate politics of a Catholic Cardinal." Thompson indicts Peccei's vision and his method as imperial, as tending toward a coercive reordering of economic and political relations under concentrated secular power, and because it is prosaic, unable to transcend the forms of thought that gave birth to the problems.[10] As Thompson puts it, "If you are going to humanize technology, you're not going to be able to do it within the limited terms of books and civilization and other older containers. You've got to go very far out."[11] In order to go very far out, Thompson has seriously proposed an effort to bring about a reunion of scientific and mystical thinking in small-scale institutions that embody a vision of the future.[12] In my view, such an orientation, although preliminary and tentative, provides solid ground upon which to build up a relevant kind of consciousness about the future receptive to knowledge and wisdom, and yet oriented toward change and decency. My only caveat is that such an enterprise may be too aloof from the urgent demands for present response. To choose residence in a monastic sanctuary entails a rejection of direct participation. Adherents become well-fed spectators in a world of hunger and bloodshed, even unwitting accomplices of those Forest Hills homedwellers who, while Kitty Genovese was killed, watched from their living room windows without even

73

lifting a phone to summon the police. So I would add that we cannot become trustworthy about the future unless we show signs of being trustworthy in the present. For Americans, at least, it is morally (and hence intellectually) impossible to propose a new Jerusalem, and yet at the same time be agnostic or indifferent about genocide and ecocide in Indochina.[13]

These comments about a new consciousness are designed to set the stage for an inquiry into the future of world order. I accept fully the mandate for a planetary, future-oriented perspective, but seek to avoid any dogmatic assumptions that its realization depends on objective or detached knowledge. In this sense, the unity of thought and feeling are essential ingredients of a relevant approach to the future. With these considerations in mind it still seems possible and desirable, indeed necessary, to propose new ways of envisioning—really re-visioning—the future so as to break the bonds of present constraints on moral and political imagination. In this sense my purpose is primarily educational, awakening the reason of men to the idea of wholeness as the basis for individual or collective sanity. Just as the biologist tends to focus on survival as a normative touchstone,[14] so the social scientist should be concerned with conditions of collective health and welfare. Seneca observed that "between public madness and that treated by doctors the only difference is that the latter suffers from disease, the former from wrong opinions,"[15] and it is this analogy that I would stress as focal to our efforts to provide the ingredients of a preferred future capable of rekindling hope, and hence, of remobilizing energies for action. Unlike Seneca, however, I believe that public madness is a result of far more complex forces than merely wrong opinions, or, put differently, that the wrong opinions rest upon cultural and philosophical underpinnings that cannot be removed merely by argument or evidence.

For instance, the American involvement in Indochina illustrates an extreme form of public madness, but there is something very functional about its persistence given the realities of American culture. Kurt Vonnegut, Jr., correctly I think, places emphasis on Americans as winners in a world of scarcity and misery: "The single religion of the Winners is a harsh interpretation of Darwinism, which argues that it is the will of the universe that only the fittest should survive."[16] Inuring itself to the fighting in Vietnam all these years provides American society with the moral hardening needed in a world of growing inequity and desperation. To quote Vonnegut once more: "The Vietnamese are impoverished farmers, far, far away. The Winners in America have had them bombed and shot day in and day out, for years on end. This is not madness or foolishness as some people have suggested. It is a way for the Winners to learn how to be pitiless." And why is this necessary given America's prowess and prosperity? Because the Winners understand "that the material resources of the planet are almost exhausted, and that pity will soon be a form of suicide."[17]

Vonnegut's perception of the psychology of the ruling group is very important because it suggests that what is public madness on one level (the waste of resources, the dissent at home, the loss of international prestige—in other words the whole gamut of arguments that have turned liberals against the Vietnam War over the years) is the essence of rulership on another.[18] And this brings us to the essence of the world-order context that exists at the present historical moment: Are the prescriptions of the Winners necessary or desirable? Are there alternatives?

Posing these questions somewhat differently: Is the Darwinian ethos adaptable to our contemporary situation? Does the notion of the survival of the fittest provide sound counsel for action on the part of power centers within the present system of world order? There is no doubt that profound chal-

75

lenges are being posed by the crowding of the planet and the depletion of its resources. These challenges are intensified by the destructive technology of warfare that is being spread throughout the world, as well as by the vulnerability of the post-industrial world to well-conceived disruption. We have a world-order crisis of unprecedented magnitude, involving issues of irreversible ecological decay by accumulating pollution, phenomena of potential mass famine and pandemics, and the possibility—never more than an hour away—of catastrophic war.

In such circumstances, we are, recalling Seneca's precept, confronted by a prime necessity to identify "wrong opinions" as quickly as possible. This necessity can be approached from many angles. We will here consider it from the viewpoint of world-order reform: What changes are desirable and possible within the next several decades? How can we most effectively think about world-order change so as to conceive of real alternatives in a manner that is the captive neither of present clichés nor prospective fantasies? We ask these questions as a social scientist respectful of evidence and discipline, yet sensitive to the criticism that "scientific method" as it has been understood by Western social scientists has excluded many realities that will inform the politics of transformation. William Irwin Thompson, Doris Lessing, or Kurt Vonnegut are world-order thinkers of relevance precisely because they are "open" to these nonrational sources of insight into the present and future.

II

Our approach to the study of world order reform is based on two orienting imperatives:
  —the methodological imperative;
  —the normative imperative.
Each of these imperatives requires some explanation.

The Methodological Imperative. We accept here the need for disciplined inquiry, for systemic comprehensiveness, and for continuous revision of our designs for the future. To be concerned with world order is to be concerned with basic relations of power and authority that operate throughout the planet. Such relations are not exhausted by an analysis of state sovereignty in the modern world, although such an analysis is of great significance. There are many actors on the world scene other than national governments that need, increasingly, to be taken into account. The activities of the multinational corporations, international institutions, transnational associations (whether of airline pilots or Red Cross officials), and of change-oriented social movements operating at all scales of organization, are important aspects of the world-order system, and condition its potentiality for change.

Of basic significance is the hypothesis that authority is related to power, and that in the world-order system of today power is concentrated in a relatively small number of large national governments and closely affiliated corporate entities. Such a hypothesis has tactical implications. It implies, for one thing, that changing the world-order system depends on altering the perceptions, values, and personality features of ruling groups in these key actors or in replacing these ruling groups with new elites having different perceptions, values, and personality features. As far as method is concerned, there is a corresponding need to concentrate on what is happening with domestic social movements, especially in the largest states, and in depicting the shape of struggle, if any, between territorially based political power and market-based economic power, or put differently, between government leaders and multinational managers.[19]

Such an emphasis reverses a traditional concern of world-order reformers with external relations between governing elites that are presumed responsive to the common interests of world society. These reformers have posited designs for new arrangements of power based on an idealistic image of peace on earth and goodwill.[20] These designs have never seemed relevant because they lacked a theory of change and were based on an ethos antithetical to the conflict-oriented ethos that follows from state sovereignty in a laissez-faire framework of centralized restraint and regulation.[21] Often these designs involved "world government" based on "law," an outlook embodied in the most influential modern variant of traditional world-order thinking.[22] World-order reform consists of convincing influential individuals that a particular set of proposals is desirable, and this persuasive strategy presupposes, competely contrary to observed fact, that the existing structure of world order is administered by reasonable men of goodwill who are susceptible to persuasive techniques of influence.

After World War I, but even more after the atomic explosions at Hiroshima and Nagasaki in World War II, the plea of world-order reformers has rested on a claim of alleged necessity. In other words, the argument for reform is backed up by an assertion that the existing system is heading for destruction given the possibilities for catastrophic war that exist at the present time.[23] Such a warning has been recently bolstered by a declaration of ecological emergency on many fronts.[24] Here again, however, it is clear that such a plea has no capacity to induce fundamental world-order reform. Indeed, the main impact of ecological pressure may be to drive the Darwinian tendencies of the present world-order system to a further extreme, namely encouraging an even more explicitly imperial arrangement with a reduction in the number of relevant political and economic centers of decision and a widening gap between winners and losers. There are various ways to respond to an objective situation of world-order danger, and it is not in any sense methodologically accurate to assume that "reform" is in a progressive direction given an idealistic outlook on man and the world. We insist, then, that the idea of future patterns of world order take account of an array of alternatives to convey a sense of choice and of process.

At the present time there is a consensus among the powerful that adjustments in the world-order system are needed to maximize the short- and middle-term interests of the strongest and richest governmental actors. The shape of this design has been spelled out by Richard Nixon and Henry Kissinger, and must be understood for what it is, world order reform engineered by and for the sake of the winners. We do not have enough evidence at the present time to demonstrate that it won't work, if specified in relation to goals associated with the maximization of the security, wealth, and influence of these dominant actors and those aligned with them for the next few decades.

The Normative Imperative. Our method of inquiry leads us to realize that there are a number of plausible models of world-order reform. Each model is capable of realization within the next several decades and each has its own immediate implications for policy, action, tactics, and belief systems. We believe that there is no objective way to demonstrate which model is preferable outside the realm of moral choice.

Given background conditions of war dangers, resource scarcity, and widespread misery, the fundamental normative choice is between an *actor orientation* and a *community orientation* with respect to world-order reform. An actor orientation means that each national center of power and wealth orients its policy toward maintaining its position of preeminence in relation to others in the system (potential rivals) by reliance on threats, force, and cunning.[25] Beyond engaging in a pietistic rhetoric—that distracts attention from the real or operational code of behavior—there is no concern for the suffering of the victims or those who are denied the fruits of participation; indeed, there is constant vigilance against challenges from the dispossessed and a ruthless willingness to repudiate their claims and repress or persecute their spokesmen. This is the world-order significance of the American role in Indochina and the Soviet willingness to accommodate their foreign policy to such outrages despite their adversary positioning in the conflict from a geopolitical perspective.

A community orientation means that the basic relationships of power and authority must become increasingly contractual and voluntary in character. Such goals mean that the problems of human existence on the planet must be approached without deference to artificial boundaries, whether of states, races, classes, or castes. This kind of problem-solving means that issues of poverty, pollution, and repression are essential concerns of world-order reformers who are guided by a vision of human community in which men live in harmony with each other and in relation to their natural habitat. This outlook can be implemented by a number of different models of world order, the choice reflecting issues of feasibility and of time horizon. There is no federalist scheme buried in the normative imperative that is built up on a community orientation. Again, as with the basic concern with reform prospects, it is helpful to explicate some alternatives and to choose from these alternatives on the basis of principled judgment. As we indicated in the introductory section, the sequel to thought is action, the relationship being well comprehended by cybernetic terminology with its emphasis on feedback loops. Every assertion of preference, whether in relation to the goals or tactics of transition, needs to be continuously reconsidered in light of developments and adapted to experiences and to changing forms of consciousness.

The permanent elements in the normative imperative are

the insistence on thinking of welfare in terms of wholes, the human species, the earth, the overall pattern of linkage between man and earth, as well as between the present and the future. It is the insistence on the coherence of the whole that makes the normative imperative compatible with the methodological imperative, and creates the basis for substantive investigations of the present prospects for world-order reform.

III

The failures of the present system of world order suggest the need for alternatives. An important contribution by specialists would involve the design of credible alternative systems of world order that have some prospect of realization within the relatively near future, say by the first decade of the twenty-first century. Such design concepts have in the past—within the wider tradition of utopography—been harmed by two basic defects:

1. No conception of transition linking the present to the future;

2. A failure to envision world-order solutions other than by the replication on a global level of the concentrations of power and authority of the sort now manifest in the governance of large sovereign states.

We propose a new social science approach to the design of future systems of world order that seeks to overcome these defects; this approach can be depicted in terms of its own set of orienting conditions.

1. The future system will itself be a step in a transitional process of political development on a global scale; in other words, processes of aspiration and projects for change are essential attributes of healthy behavior and parallel biological processes of evolution; there is no final solution of the world-

order problem, but only a series of transitional solutions; each achieved utopia generates its own new horizon of aspiration;

2. The search for a new system of world order depends upon mobilizing support for a series of explicit values; world-order change—solving the transition problem—depends on achieving and implementing a normative consensus;

3. This normative consensus is animated by an overriding concern with initiating a humane process of adjustment to dangers associated with problems of ecological imbalance arising from crowding, depletion of resources, and poisoning of the biosphere; as Jay Forrester has correctly suggested, "Civilization is in a transition zone between past exponential growth and some future form of equilibrium."[26] The focus of inquiry, then, should be directed toward *dynamic equilibrium* models of world order that can be achieved without undue trauma and that can be sustained without recourse to repression; these ethical dimensions of world order relate to attitudes toward violence, satisfaction of basic human needs, and social and political conditions that are compatible with a sense of human equality and with an affirmation of the worth of the individual human being.

In this chapter we can give only a sense of direction.[27] Figures 1 and 2 attempt to set forth the basic framework of thought that has been developed in response to the five conditions laid down and discussed in the preceding paragraphs.

Figure 1 depicts the basic set of concepts that seem useful in developing a systematic method of thinking about future systems of world order (i.e., S_1, S_2 ... S_n) and of conceiving the transition path from $S_o - - - \rightarrow S_1$ (i.e., either in analytic terms of t_1, t_2, t_3 or in temporal terms of $t_{1970's}$, $t_{1980's}$, $t_{1990's}$). Figure 2 suggests one line of transition in relation to value priorities of peacefulness, social and economic welfare, environmental quality, and human dignity. This transition path emphasizes several sequences of development that appear to

82

System (S) Level (S_0 = S at Origin)	$S_{-1} \longleftarrow - - - - - - - - S_0 - - - - - - - - - \longrightarrow S_1 - - - - - - - - - \longrightarrow S_2$
System Level (chronological subscripts)	$S_{1914} \longleftarrow - - - - - - S_{1973} - - - - - \longrightarrow S_{2000} - - - - - \longrightarrow S_{2050}$
Transitional Stages for the interval $S_0 - - \rightarrow S_1 - - \rightarrow S_2$	$S_0 - - - - - \longrightarrow S_1 - - - - - \longrightarrow S_2$ $\quad (t_1)\,(t_2)\,(t_3) \qquad (t_4)\,(t_5)\,(t_6)$
Transitional Stages for the interval $S_{1973} - - \rightarrow S_{2000}$ with chronological subscripts	$S_{1973} - - - - - - \longrightarrow S_{2000}$ $\quad (t_{1973})\,(t_{1980})\,(t_{1990})$

Figure 1.

be preconditions to the emergence of S_1 (i.e., on the level of action the correlation between consciousness and t_1 or $t_{1970\text{'s}}$, between mobilization and t_2 or $t_{1980\text{'s}}$, and between transformation and t_3 or $t_{1990\text{'s}}$; on the level of primary institutional arenas a similar sequence of correlations with t intervals is depicted). The basic purpose of Figure 2 is to embody a conception of global change that will have to accompany any serious process of designing and achieving a new system of world order. The number of stars in each box signifies additional emphasis that builds upon the achievements of the prior interval. Thus, by the end of t_2 both war and social and economic welfare considerations have six units of cumulative change. The distribution of stars is meant only to be a rough approximation of relative degrees of effort and achievement at the various stages of the transition process. It should be understood that this is a *conception*, or at most a *prescription*, rather than a *prediction*. In terms of our conception, it is possible, even likely, that t_1 will never come to an end; we are merely proposing that a way to move from $S_0 - - - \rightarrow S_1$ is to proceed along this path of

83

Transition Path S -------> S_1

Problem Focus Change Orientation Institutional Focus	War Consciousness Domestic Arena	Poverty Mobilization Transnational and Regional Arenas	Pollution Transformation Global Arena	Human Rights Transformation Global Arena
Transition Stages / Temporal Subscripts / Analytic Stages				
t_1 $t_{1970's}$	*****	**	*	
t_2 $t_{1980's}$	**	****	***	*
t_3 $t_{1990's}$	*	***	****	***

[N.B. The number of stars in each box is roughly proportional to the degree of incremental emphasis in each t interval.]

Figure 2.

Temporal Interval	Positive Goal	Creativity	Self-realization	Joy
$t_{2010's}$ $2020's$	t_4	***	**	*
$t_{2030's}$	t_5	**	***	**
$t_{2040's}$	t_6	*	*	***

Figure 3.

sequenced transition. The transition process proposed is but one of an infinite series; alternative transition paths $S_0 ---\rightarrow S_1$ can and should be studied as part of a broader inquiry into comparative systems of world order.

To make the point clearer that S_1 is itself a transitional solution to the challenge of world order, we depict in Figure 3 the subsequent process of transition from $S_1 ---\rightarrow S_2$ with a similar suggestive profile of sequenced transition.

Figure 3 merely suggests the more personalist follow-on to a successful effort to save the planet from misery and extreme danger. In a sense $S_0 ---\rightarrow S_1$ is concerned with a world-order rescue mission, whereas $S_1 ---\rightarrow S_2$ is concerned with the positive task of building a system of world order that fulfills human potentialities for growth and satisfaction.

The only element that has not yet been portrayed in our approach is some conception of the structural character of the outcome of transition. Our focus is upon the basic arrangement of power and authority as an organizing energy in planetary affairs. Again, the objective here is to suggest a *mode* of thinking rather than to argue on behalf of a particular configuration of power and authority. At the same time we would like to set forth models of world order that correspond with various relevant lines of real-world preference and prediction. In particular, we would like to offer four kinds of models of world order based on an acceptance of *dynamic equilibrium* as the fundamental prerequisite:

—Nixon-Kissinger (Figure 4)
—Doomsday (Figures 5 and 6)
—World Government (Figure 7)
—S_o – – – $\rightarrow$ S_1 (Figure 8)

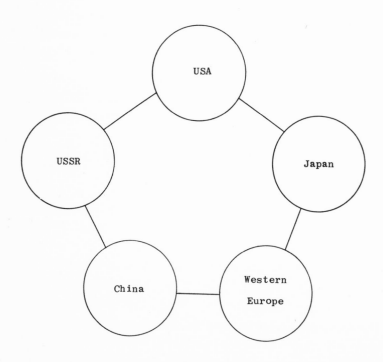

Figure 4.

The basic attribute of the Nixon-Kissinger model of world-order reform is the notion that existing centers of industrial capability, and hence military potential, are capable of providing a stable and acceptable system of world order for the indefinite future. This system rests on moderation of means and ends among the five dominant actors who work out patterns for efficient cooperation and limited competition, based on a general acceptance of the geopolitical and geoeconomic status quo. The Nixon-Kissinger model accords little emphasis to social and economic justice and it is pre-ecological to the extent that it assumes the capacity of competing national units to behave in a manner compatible with the maintenance of ecological balance. A concert of dominant actors may avert short-term breakdown in S_o, but it hardly provides a satisfactory response to either the negative threats posed by the present crisis, nor does such a world-order design offer any prospect of fulfilling the positive potential for human social and individual development.

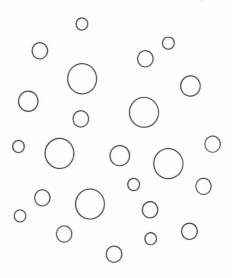

Figure 5.

Figure 5 depicts the breakdown of the present world-order system (S_0) into a series of gravely weakened and uncoordinated units. These units may or may not possess a statelike formal identity, but there will be a loss of internal cohesion and a virtual collapse of any capacity to enter into stable external relationships. The state system will have disintegrated into a condition of world anarchy or chaos, and political organization of any comprehensive kind will not exist. Poverty, violence, disease will be rampant, and there will not be any prospects for recreating conditions of order and justice that are even comparable to S_0.

Figure 6 illustrates the other main line of response to the collapse of S_0. Here remaining power is concentrated at a single focal point. Planetary resources are exceedingly scarce relative to human needs, and privation is widespread. A small elite runs the world in a highly dictatorial, repressive fashion. No opposition is tolerated. The system of world order resembles a police state of the sort now associated with the worst national tyrannies.

In effect, the collapse of S_0 is likely either to accentuate

Figure 6.

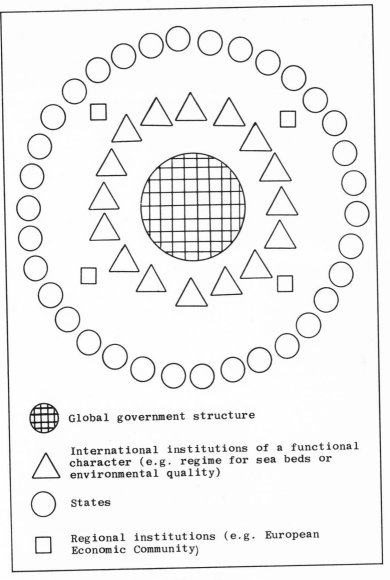

Global government structure

International institutions of a functional character (e.g. regime for sea beds or environmental quality)

States

Regional institutions (e.g. European Economic Community)

Figure 7.

89

centrifugal tendencies (in which case Figure 5) or centripetal tendencies (in which case Figure 6). In either eventuality, the present world-order system with all its deficiencies would itself look like a utopia by comparison. It is important to understand that regression, as well as positive change, is possible in relation to the future of world order.

Figure 7 depicts in very crude terms a world-federalist response to the inadequacies of S_o. Some constitutional conception of this sort has dominated the visionary literature of world order for centuries. The notion of world government seems to satisfy the basic human craving for unity and order. As such, it captures something very fundamental about the basic direction of world-order reform. At the present stage of international relations, however, a world government solution does not seem attainable except as a response to a doomsday situation, in which case it is likely to be a dysutopia of the sort projected in Figure 6. That is, there is no credible transition path that can be followed over the next several decades that could lead reliably toward world government of a benevolent character.

In addition, governmental solutions to human problems have not often worked with success at a national level. The prospect of a huge global bureaucratic presence, especially if combined with the sort of technological apparatus that will be available, does create inherent reasons to be wary of a world government sequel to S_o even if it were attainable. These reasons include problems associated with domination, conformism, and excessive administration, as well as concern about such a concentration of military capabilities and political powers at this stage in the growth of human consciousness.

Figure 8 represents a compromise between the state system and the sort of world government solution projected in Figure 7. $S_o - - - \rightarrow S_1$ has to be conceived in relation to a systematic study of transition prospects, tactics, and sequences as sug-

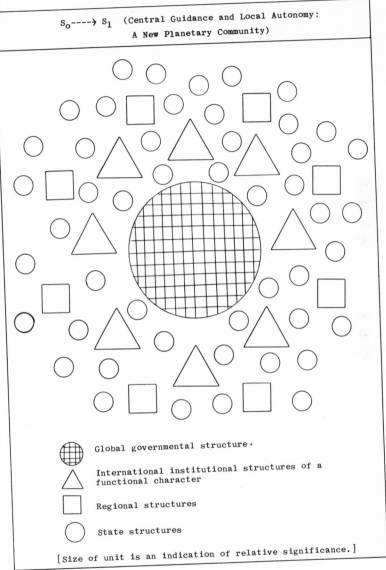

S$_0$----> S$_1$ (Central Guidance and Local Autonomy: A New Planetary Community)

Global governmental structure.

International institutional structures of a functional character

Regional structures

State structures

[Size of unit is an indication of relative significance.]

Figure 8.

gested in Figures 1 and 2.[28] The basic constitutional principles embodied in the profile of S_1 are to achieve maximum co-ordination and guidance on the basis of minimum coercion and bureaucratic intrusion. Great emphasis is placed on diminishing the existing power apparatus in the superstates (rather than in transferring power upwards) and in designing into the structure numerous checks and balances, as well as upward avenues of influence and participation.[29] Figure 8 does not indicate the presence of transnational groupings, which we believe will be very important in S_1, given especially the altered nature of political consciousness, which is a precondition for $t_1 ---\rightarrow t_2$. We believe that transnational economic, cultural, and political groupings will enjoy great importance in S_1, and indeed may provide an emerging focus for human loyalty displacing gradually the symbolism of national patriotism.

I V

In Section III we set forth the skeletal elements of a way of thinking about the future of world order. This way of thinking is shaped by a normative orientation toward what is desirable in human relations and political arrangements. Such an approach amounts to a plea for restructuring inquiry into international relations in the direction of a newly conceived academic discipline of *comparative systems of world order*.

—*comparative* in relation to past, present, and possible future arrangements of power and authority in human affairs;

—*systemic* in the sense of being as rigorous as possible with respect to the totality of behavior embraced by the subject matter;

—*world order* in the sense of a normative orientation toward the appraisal of the performance or desirability of a particular global authority; normative both in relation to tasks performed (e.g., human needs, peace, conservation of resources, etc.) and in relation to responsible aspirational projects (i.e., conditioning goals by feasibility studies of transition prospects).

On such an intellectual basis a new conception of world order can begin to emerge that is sensitive to the role of thought and reason in accomplishing an essential reunion between feeling and action. The splits between thought and feeling on one side and between thought and action on the other have often rendered analysis of world order sterile, and conveyed either an impression of intellectual aridity or one of political futility.

PART TWO

THEORY AND METATHEORY:
NORMS, METHODS,
AND PRESUPPOSITIONS

5

The Twilight of Physical Descriptions and the Ascent of Normative Models

1. The Myth of Pure Descriptions

The sharp separation of the descriptive from the normative is the heritage of modern empiricism. Classical empiricism is a philosophical justification of the world view based on physical science. This world view holds that the world is predominantly empirical and can be best rendered by physical science, whose ultimate conceptual components are *descriptions*, or scientific explanations. Within the consistent scientific world view, norms and values are relegated to limbo. They have no genuine cognitive status. The preponderance of physical descriptions within the world view based on physical science is by no means accidental. The ideology of modern science, which itself is a value system, so interprets other value systems as to make them either unimportant or meaningless.

Let us first of all notice that description of what is *out there* largely depends on what we *assume* to be there. In themselves, descriptions are either meaningless or inherently ambiguous. Out of the context to which they belong, or out of the system of which they are a part, descriptions are unintelligible. Karl Popper used to ask his students to "observe and describe." Students were invariably perplexed and unable to follow this simple request. Their inevitable question was: Observe *what*? We always have to observe *something* in order to determine something. The myth of pure observation is a part of the Baconian heritage. We have now reconciled ourselves to the fact that there is no such thing as pure observation. Every observation is a *directed* observation, is an observation for, or against, a point of view. Every mind is a "contaminated mind," a mind constructed of a network of suppositions and assumptions. This is no less true of a bushman than of a nuclear physicist. Each works within a network of presuppositions which make him see and *describe* the reality of things surrounding him in a specific way. To use Kuhn's language: All descriptions are paradigm-conditioned—not only scientific descriptions, but magical descriptions as well.

More enlightened philosophers of science accept nowadays that all observation is theory-laden.[1] I should like to suggest that *all description is Weltanschauung infused*, is a description which renders to us not only "things as they are," but also renders a part of the world view through which "things as they are" are construed and comprehended. A description in itself is inherently ambiguous. We cannot *just* describe. All description is bound by a particular ontology, or a particular world view. A given set of descriptions is conditioned by a world view, and it in turn articulates and exemplifies this world view. Physical descriptions give us an account of the physical world. Aesthetic descriptions give us an account of the world of art. If we insist that the world is exhausted by the sum total of physical descriptions and that other descriptions do not matter, we subscribe to empiricism, or a form of positivism. If we insist that aesthetic descriptions, and in general the descriptions of spiritual and cultural aspects of man's life, are equally valid, we go beyond the bounds of empiricism.

As I said, through our descriptions we render an aspect of a world view. From which it follows that all observation is not only theory-laden, but also *Weltanschauung* infused. A command to observe certain things, in order to elicit certain answers, is a command to describe them. Observation without description is a psychological possibility, but not a conceptual one. We can imagine it as a nonlinguistic mental state, but we cannot bring it to the level of conscious thought because then it immediately *becomes* a description. The "mere" act of observation, if it is made conscious and articulate, is an act of description, an act of rendering a part of the world view by which this observation and description is conditioned and determined.

Let us consider some examples to further illustrate the dependence of descriptions on the a priori assumed world view. Let us take a classical behaviorist of the Skinnerian type on the one hand, and a psychoanalyst of the Freudian school on the other. Let us confront them with a mentally disturbed patient. How will they describe the behavior of the patient? In vastly different terms. The same behavior is described in such incompatible ways! How is it possible? And who is right, the behaviorist or the psychoanalyst? Now, it is not quite true that the *same* behavior gives rise to such different descriptions. The *observed* behavior is *different* in each case. What each investigator observes is a function of the conceptual framework which he imposes on the phenomena he examines. Thus, the apparent sameness of the behavior they study will be resolved in two different sets of categories and concepts. Indeed, the two different sets of categories and concepts with which they approach the subject will lead them respectively to construe different pictures of the seemingly same behavior. Is it not the case, then, that they describe two different species of behavior? And the very term "behavior," to begin with, will denote for them quite different things. Let me explain.

The psychoanalyst observing certain "odd" responses of the patient is not merely observing, but at the same time reconstructing the possible psychic causes of these "odd" responses, and might even conjecture what kind of experiences in the childhood of the patient might have been at the root of his illness. The behaviorist observing certain "odd" responses of the patient is also not merely observing (there is no such thing as *mere* observation), but is playing with conjectures about what kind of positive-reinforcement therapy could remove those undesirable "odd" responses.

For the behaviorist, the "psychic causes" of the psycho-analyst form an indulgence in fiction. For the psychoanalyst, the positive-reinforcement therapy of the behaviorist is a vulgar manipulation of the human being without understanding what persons are. Each respective conceptual framework controls the *process* of observation and the *outcome* of description, as well as the *judgment* cast on the rival description.

The validity of descriptions, as I have said, is relative to the conceptual framework, or to the world view which we adopt. How do we adopt a conceptual framework, or a world view? Here is the rub. Usually we do not. It is thrust upon us from early childhood, or imposed on us in a subtle way during our school years. Poincaré, Duhem, Ajdukiewicz, Quine, and others have shown that a change of the conceptual system, or the conceptual apparatus, may lead to a different world picture, which means to alternative descriptions. These thinkers insist that we have a large latitude of choice regarding what conceptual framework to adopt. There are no compelling or necessary reasons to make us adopt this framework rather than the other.

When the various conventionalists, such as Poincaré and Ajdukiewicz, insist that we have a large latitude of choice in selecting a conceptual framework within which to describe phenomena, they do not really mean choices in the global sense, but rather choices within the accepted culture and the accepted (scientific) world view. Conceptual strategies developed by conventionalists were designed to overcome the intellectual crisis of the late nineteenth and the early twentieth century. At that time classical physics, classical geometry, and traditional logic were undermined by new extensions of knowledge. Conceptual readjustments, sometimes quite considerable, had to be made in order to accommodate new insights. In short, conventionalist adjustments of the scope of human knowledge did not question or challenge the supremacy of

Western man, nor did they challenge the supremacy of the scientific world view. Conventionalists deliberately limited themselves to a certain world view, to a specific culture with its underlying values. But we can follow their insight beyond the limits to which they dared go.

The vital lesson of conventionalism is that our knowledge at least partly constitutes our world. With a different knowledge we, so to speak, receive a different world. We have more than one choice of knowledge.[2] The choice of knowledge is like the choice of values—we are not limited to one choice only. The choice of values is ultimately justified by the long-term service these values render to the species. And so it is with knowledge. In a sense a choice of knowledge *is* a choice of value.

2. The Brittleness of Physical Descriptions

To describe the world is not only to describe its most apparent physical furniture. We have so refined the language of our descriptions that it handles physical descriptions masterfully and sometimes even superbly. Man's existence is sustained by physical knowledge, but not exhausted by it. We have made a mistake, sometime in the seventeenth century, by assuming that all descriptions can be reduced to physical descriptions. Physical descriptions impress us because they are easy to make and easy to test. In its excessive preoccupation with physical descriptions, Western civilization betrays its shallowness, its inability to probe the deeper aspects of man's universe, its smug satisfaction with appearances. We are a facile civilization, sparkling on the surface, shallow inside. It is imperative to break the spurious dichotomy between the descriptive and the normative, established by prophets of shallow empiricism, in order to bring back to man his due depth, in order to bring back to knowledge the ignored aspects of reality, in order to bring back to descriptions the wealth and depth of which they have been robbed as the result of the mindless pursuit of empiricism and positivism.

Our civilization, as epitomized by scientific knowledge, has simplified, if not vulgarized, the variety of our interactions with nature and cosmos. During the course of its development modern science has consistently attempted to purify the universe of nonmechanistic elements. This simplification of the world picture and the conceptual clarity that ensued is a stupendous achievement of the human mind. But it has also turned out to be an enormous limitation. We have so "purified" the furniture of the world, and so atomized it into smaller and smaller components, that we have lost the grasp of the essential wholes through which nature operates and through which life is lived.

In this context we can clearly see the pseudoneutrality of science, its observations and descriptions. Science is not neutral and its descriptions are not exact or faithful renderings of things "as they are." Science contains in itself a historically developed ontology, for it tells us what is there to be seen and explored. It contains a historically developed methodology, for it tells us what method to use and how to explore what is there. And furthermore, science contains a historically developed eschatology, for it tells us to what purposes we are exploring, it tells us indirectly that progress to be achieved through science will lead to the fulfillment of man on earth, and perhaps even to a secular salvation. These three components which are built into the edifice of science—ontology, methodology, and eschatology—are not sufficiently justified. Yet they control the nature of the world we experience through science, and they also control the character of the descriptions science renders.

What is at stake now is not a semantic or even an epistemological exercise consisting in the redrawing of the boundaries between the descriptive and the normative, but rather a radical change of the whole idiom of present knowledge. In spite of our analytical and descriptive dexterity, we find ourselves stupefied and dumbfounded by our social, human, and ecological problems, all of which require the recognition of phenomena, and thus the recognition of descriptions, which go beyond the scope of empiricism. We thus witness a pathetic spectacle—a huge army of dexterous humans juggling with words and formulas that have no bearing on, or relevance for, the reality that is at the root of our problems. The inbuilt limitations of physical descriptions, limitations which have pervaded our language and thinking, not only give us a distorted view of reality, but also, and above all, bar us from alternative descriptions and alternative solutions; at present these limitations are like a cloud hanging over our horizon and preventing us from seeing more clearly and more penetratingly. From Forrester to Skinner a great many efforts to find effective solutions to our global problems are hopelessly bogged down in their inception because of the limitations of physical descriptions which are still recognized as the criterion of validity in our descriptions.

Thus, new knowledge based on different kinds of descriptions is not only an epistemological imperative, it is also a social imperative: it is an imperative for our survival. What we are seeking, without perhaps being fully aware of it, is not so much improved science, or more science, but a different idiom for living, a different idiom for our interaction with nature and cosmos. We must liberate ourselves from the pernicious assumption that present Western rationality and present Western science are the alpha and omega of all knowledge.

But, the opponent might argue, science is a quest for truth and the discovery of truth has great survival value. If we hit upon good guesses, good conjectures, and good theories, we survive. If our guesses and theories happen to be false, we perish. Thus truth is tantamount to survival; and falsity to extinction.

The evolutionary biologist G. G. Simpson has argued that there must be some parallelism between the features of the world and our perception of it, that our knowledge of the world must be valid to a large degree, otherwise we would not have survived. If the monkey, he says, did not have an accurate perception and understanding of spatial relationships, the next time it jumped from one tree to another it would be a dead monkey.

This is quite true. But let us notice that the monkey does not possess scientific knowledge upon which to rely in performing its superbly accurate jumps. No doubt animals do possess knowledge that enables them to solve their countless problems. But this is not scientific knowledge. Knowledge is necessary for the survival of any species; but not necessarily *scientific* knowledge. Man, as a species, has long "survived" without scientific knowledge. With scientific knowledge our short-run survival potential has increased; or so we thought at least. We now question the long-run survival potential of scientific knowledge.

Derek Price has estimated[3] that about 85–90 percent of all scientists that ever lived are alive today. The volume of knowledge (one would perhaps prefer to say—of scientific information) is supposed to be doubling every ten years. But to what effect?

Dinosaurs have perished because of their sheer size. Is it not the case that the growing volume of factual (scientific) knowledge is choking us? If so, then the increase in the *volume* of scientific knowledge does not necessarily mean an increase of its *value* for the human species. We seem to have reached the point of diminishing returns within the present idiom of scientific knowledge. Unintended detrimental consequences of science and scientific technology (which is, after all, based on science) in terms of environmental pollution, as well as human pollution, do not seem to justify our total commitment to present science.

Scientific descriptions are quite useless for expressing a great deal of the content of human experience. Moreover, there are certain claims, assertions, and beliefs held by people which, if measured by the standard of scientific veracity, would have to be recognized as plainly false. And yet, some of these apparently false beliefs have great survival value. Ideologies that bind people together, myths that have inspired and sustained people (not only in antiquity, but in modern times), aspirations and claims of artists to have grasped the ultimate reality, and last but not least the claims of religion to have revealed ultimate truths, if scrutinized under the scientific microscope, would all be shown, at least to a large degree, to be based on illusion and falsehood. And yet these apparent "falsehoods" sustain us and seem inseparable from the human condition. In short, man's culture, man's values, man's moral codes are a supreme achievement of the human species. They cannot be abandoned without reducing the species to the level of lower animals, and yet they cannot be subsumed under the umbrella of scientific reason.

I take exception here to Karl Popper, who insists that if we do not follow the path of truth, we perish. On the contrary, if truth is identified with scientific truth, then if we do not depart from scientific truth, indeed if we do not embark on the path of "falsehood," *then* we perish. By "falsehood" I mean all kinds of beliefs, assertions, and assumptions that are embedded in our myths, ideologies, and, sometimes, in moral codes and social institutions which we cherish, act upon, and are sustained by, but which nevertheless can be shown to be incompatible with scientific truths. For a long time we have disregarded this glaring chasm, hoping, in the depth of our scientific reason, that more refined science will be a substitute for philosophy, art, and religion. We cannot cherish this hope any longer. Science has been changing and is changing. But the change that awaits it is of a Copernican dimension: not a little refinement, but a radical revolution.

Present science constantly attempts to reduce richness and variety to simplicity and homogeneity. This reductionism of science has paid handsome dividends. As a methodological program, reductionism has been more than justified. Enormous advances have been made by a deliberate and conscious adoption of the reductionist attitude, by deliberate attempts to explain phenomena of higher levels by the properties of the phenomena of lower levels. But the methodological reductionism and the descriptions it renders cannot be detached from a larger world view with which they are connected. We may argue however that one can be a methodological reductionist in the laboratory—trying as hard as he can to reduce the complex phenomena of life, for instance, to the level of molecular biology—and yet he does not have to succumb to the epistemological, or the ontological, reductionism. That is to say, he does not have to acknowledge that because he is a methodological reductionist he must reduce *all knowledge* to the concepts and categories of lower levels; this would be epistemological reductionism. Nor does he have to admit that all the furniture of the world must be reduced to physico-chemical elements; this would be ontological reductionism. Logically, indeed, one type of reductionism does not imply the other. Yet, the practice of methodological reductionism may have insidious effects on the human mind. V. E. Frankl has argued, in an essay tellingly entitled, "Reductionism and Nihilism,"[4] that reductionism today is a mask for nihilism. Professor of psychiatry at the University of Vienna, and a practicing therapist, Frankl suggests that the widespread feeling of inner emptiness among people, and the consequent overflowing of clinics and consulting rooms, is the direct result of the modern scientific world view in which values are denied. Since science and its techniques are largely reductionist, the inevitable conclusion appears to be, Frankl argues, that reductionism is the only philosophy in which one can believe. Since this philosophy cannot sustain man's quest for values and meaning, all kinds of psychological and psychiatric problems result.

This conclusion does not necessarily follow. In actual fact we can *separate* methodological reductionism, as the mode of our behavior in the laboratory, from other forms of our behavior elsewhere. But this kind of separation only illustrates the favorite presumption (or should we say, an unwarranted dogma) of scientists that the world is made of separate and disconnected bits, and so is human behavior. Science endlessly separates, atomizes, detaches, abstracts; whereas the life of animals, of societies, and of human individuals is lived in wholes. And once again we must bear in mind that descriptions are never things in a vacuum: they are always an aspect of a larger world they represent. Accept descriptions of a certain kind and you are involuntarily accepting a part of a larger framework within which these descriptions make sense. It is for this reason that methodological reductionism is so pernicious.

Frankl's main message seems to be that when thousands and thousands of young students are exposed to an indoctrination along reductionist lines, when they are taught a reductionist concept of man (man but a machine) and a reductionist view of life (life but a physicochemical process), then this is bound to have, and, in fact does have, pernicious results on their lives as human beings.

3. The Rise of Normative Models

All living systems are normative. The norms of survival and well-being of organisms cannot be deduced from the descriptions of these organisms in terms of physics and chemistry. Just the opposite, the arrangement of lower organs and ultimately of the cells and their chemistry is subordinated to these norms of survival. This is not to say that some nonmaterial entities called "norms of survival" arrange the chemistry of the cells. But rather that the telenomic character of organisms, as seen through the norms of survival, is a defining characteristic of living organisms *as living*.

The normative character of human life is even more apparent. Deprived of the skeleton of values and other norms characteristic of human existence, we end up, as Frankl puts it, with an existential vacuum and nihilism. General systems theory has been from the start an attempt to overcome the limitations of purely descriptive approaches, particularly physical models. When we examine some of the models developed within general systems theory—of von Bertalanffy and others —we realize that their success has been partly due to the fact that they are normative models. The rise of normative models signifies the reassertion of quality over quantity, signifies the dawn of the new epoch in which values are uplifted from a limbo to the level of real importance.

By normative models I mean such systems of descriptions in which values are admitted explicitly or implicitly as components of the system. Indeed, values can even become the central coordinates of the system around which all other components are built. And this central coordinate may be visible or hidden.

A model can be *explicitly* normative. It then: a) admits values as its components; b) is oriented towards specific goals.

A model can be *implicitly* normative. It is then: a) laden with values; b) its conclusions point to certain goals.

We should not fear that the admission of values will result in cognitive nihilism or total subjectivism. The values of man show both an extraordinary diversity and an extraordinary convergence. Normative models, around which comprehensive world systems are to be built, will include mostly convergent values, not divergent ones. The justification of the universality of some fundamental human values, as Ralph Burhoe[5] and others have shown, is not as difficult, let alone impossible, as we once thought.

Knowledge is a supreme instrument in the ascent of man and in the process of the perfectibility of man. We should not shun the use of the term "perfectibility" in relation to the human species, for man is a self-perfecting animal par excellence.

All knowledge must serve the human species and is only justified insofar as it aids the species in significant ways. Knowledge which does not aid the species in the process of overall survival and which does not contribute to the betterment of man is defective knowledge. It seems that the more refined scientific knowledge has grown, the more remote it has become from the real concerns of man's life. We need new kinds of descriptions in order to bring knowledge back to the human context and to human concerns.

The variety of different types of descriptions is the acknowledgment of the versatility of man's world views—which the descriptions represent. It would be quite arbitrary and indeed gratuitous to rule out all these types of descriptions as null and void except for scientific descriptions. Which brings me to a main point. We do not need to juxtapose descriptive models and normative ones. *What we need is an enlarged view of description.* In other words, we shall not have to abolish descriptions, or exorcise them as evil, but rather enlarge their scope sufficiently, so that they include the variety of elements which are necessary for an *adequate* description of the human world. Let us observe, first of all, that descriptions as free of values are mere fictions. There are no such things. Michael Scriven[6] and others have shown this quite conclusively.

111

Our normative models *are* descriptive, if we treat our descriptions as they deserve to be treated: as linguistic renderings of not only physical states, but also of social states, of aesthetic states, and other states and affairs which we generate in our interactions with the world and other people. The ascent of normative models is really the recognition of the inadequacy of previous descriptions and an attempt to provide richer, more versatile, and more adequate descriptions. If the stubborn followers of the scientific paradigm wish to maintain that the term "description" is to be reserved for scientific descriptions, then we have no choice but to talk about normative descriptions and normative models.

What we thus need and shall have to develop are new descriptive models as new modes of knowledge. In these models, or in these modes of knowledge, description will be coordinated with value. We shall look at these new descriptions not just as accurate or inaccurate, but as useful or useless from a human point of view; alternatively as life-enhancing or life-suppressing.

The outcome of this argument decisively suggests that we shall have to develop a new nonempiricist epistemology. This new epistemology will be based on the principle of empathy with the universe rather than on the principle of analytical detachment and clinical objectivity. This new epistemology will have to be based on a different kind of understanding, the understanding which goes into the depth of things rather than considering the myriad of atomistic particles into which living things can be reduced. This new epistemology will be based on the recognition that cognitive processes are not value free, and that we should therefore concentrate on those processes which are life-enhancing in the long run. This new epistemology will maintain, against Hume and all his empiricist followers, that only through the normative can we really comprehend the descriptive.

Håkan Törnebohm has persuasively argued that there is an interdependence between

| World Picture | ⇄ | Ideals of Science |

One reinforces the other.[7] Albert Wilson has convincingly demonstrated that the present world picture, constructed under the auspices of physical science, is altogether too constraining for man.[8] The inevitable conclusion that follows is that we shall have to change the ideals of science, which means changing the present idiom of scientific knowledge. For what we look for, to say it once more, is an adequate model for our interaction with the cosmos. Present science provides *a* model for such an interaction; but not an adequate one because a great variety of aspects of man's existence do not find adequate expression in it.

Adequate world system models cannot be created in a vacuum, or as the result of sheer speculation. Various movements, groups, and individuals developing soft technologies, or alternative technologies, are in the vanguard of this new kind of modeling. It should come as no surprise that alternative technologies are developed mainly by young people, who not only seek new forms of knowledge, beneficial to humanity at large, but also seek alternative societies, and constantly re-examine and reassert the primacy of human values. A viable world system model must meet the challenge and test of present social realities. For this reason it must be truly comprehensive and normative. (Read: descriptive in an enlarged sense of the term.)

Let me say a few words about one such system, which is being created and tested in actual life. I have here in mind the "Arcologies" of Paolo Soleri. Arcology is a fusion of architecture and ecology. One of these arcologies, called Arcosanti, is being built in the middle of the Arizona desert. It is a city based on entirely new principles. It is in fact an alternative society. A new environment is created to enhance man. The physical order of Arcosanti follows from its normative order.[9] Certain values and norms concerning the ends of man's life form the matrix of the whole physical system. Arcology, in short, is a world system model. (I am particularly partial to this world system model because I happen to be physically and intellectually engaged in its construction.)

The world system model that attempts to grasp the complexity of man in this complex universe, indeed, that attempts to find for man a path of survival amidst the backfiring elements, will have to be normative. Human existence is woven into a matrix of values. After a journey of fantastic exploits into the nature of matter and the inanimate world during the last three centuries, it is time to return to the sanctuary of man. Socrates is said to have shifted the focus from cosmology to man. Such a shift is required now. Socrates, remember, did not surrender inquiry into the beauty of mathematical relationships or the nature of physical things, but bent it all to enhance man.

4. Forrester's Model

Forrester's world system model is a considerable achievement of comprehensive thinking. Yet it did not escape the syndrome of reductionism via physical descriptions. Ervin Laszlo has beautifully exhibited the assumptions, limitations, and oversimplifications of Forrester's model.[10] Forrester wants to have it both ways: to grasp normative goals in terms of physical parameters. This approach has great merits, for it can be persuasive even for the hard-headed scientist. But its weaknesses are only too apparent, as Laszlo has shown. The model is only seemingly descriptive and objective, for it is laden with values, which control the choice of parameters and the outcome of our modeling. The conclusions of Forrester's model are also not innocent of prescriptions. Thus, it is a normative model, though implicitly so. By taking the appearance of a descriptive model, and by trying to keep values at bay, as it were, it denies itself the opportunity of being truly adequate for the problem with which it is concerned: the quality of life and the quality of the environment. These are inseparable from human values, and can ultimately be resolved only within an appropriate context of values.

This neglect of norms and values is glaring when Forrester attempts to discuss models as social systems. The terms "social," "society," "social values" are not much more than verbal ornaments in Forrester's parlance. Forrester *mentions* these terms many times, but never really discusses them, for these concepts are foreign to the language and capabilities of his model, determined and defined as it is by physico-economic parameters.

Though I am sympathetic to Forrester's undertaking, I am far from convinced that the road he has chosen can lead him to the goals he aims at—an adequate world system model. It appears to me that all the limitations of empiricism are built into Forrester's model. If such is the case, then the whole undertaking may be fundamentally misconceived. For all its fortes and flexibility, Forrester's model is rigid—a deterministic box governed by five physical variables. Karl Popper has argued in a beautiful essay entitled "Of Clouds and Clocks" that in order to understand the world we must imagine that it is like a cloud rather than like a clock. Forrester, so we are entitled to judge from his model, makes exactly the opposite assumption, that the world is like a clock. Forrester wants the whole world to comply with the features of his model. But it must be the other way around: his model must comply with the features of the world. And we have no reason to believe that the world is just a clock. Laplace's dream was nothing but a dream.

In his impetuosity to make his model universal, Forrester makes some extravagant claims such as:

(i)   The mind is a model;
(ii) A mental image is a model;
(iii) All our decisions are taken on the basis of models;
(iv) All our laws are passed on the basis of models.

To begin with (i), the mind is a mysterious agency and we have not yet been able to reconstruct it through models we create. The architecture and the function of the mind is so complex, so subtle, and so intricate that our models, including computers, do not even begin to approximate it. The mind as a cognitive faculty is constantly open ended, elastic, capable of operating on various levels of hierarchy, capable of switching from one order to another, capable of reevaluating its own activity.[11] Thus, the mind in its *actual* performance does *not* comply with Forrester's notion of the model. We must make models resemble the actual architecture and function of the mind rather than make the mind resemble our simplistic models.

(ii) A mental image is not a model. A mental image of a blue sky is not a model. A great deal is required of a mental image to *become* a model. And does it make sense at all to talk about mental images as models?

(iii) and (iv) are equally questionable. Our decisions and laws are often made on the basis of obscure and selfish reasons, under the influence of impulses or fear. To call all these things models is to degrade the notion of the model. Modeling is a sophisticated human activity and we should not expect to find it everywhere. But I do not wish to quibble over words, but rather to return to the central point.

Models, to my mind, are sophisticated cognitive structures. Their purpose is to aid our understanding. Models do not isomorphize the world in a one-to-one way. They are not

116

mirrors, or photographic cameras by which we can capture the world *as it is*. Rather, they are a set of filters through which we apprehend the world and render it in a specific way. No model is accurate or unbiased. All models are in the final analysis metaphors. They are flexible clouds, or at any rate ought to be, and not rigid deterministic boxes. Thus adequate world models must themselves be open-ended, dynamic, behaving like a kaleidoscope, not as a clock. If what I have said about models is correct, then Forrester has a long way to go. He does admit himself that what he proposes is the first beginning. The beginning is quite impressive. And yet one is left with misgivings. This model may be "the best thing we have" for tackling global environmental problems. But "the best we have" may turn out to be thoroughly inadequate.

Pythagoras believed that number holds the sway over flux. But for Pythagoras number was something mystical, not just a part of this "objective system of counting." Galileo resurrected Pythagoras's belief and asserted that the book of nature is written in the language of mathematics. His followers took it to mean that whatever *is*, is quantifiable and physical in nature. Of late we have come to question the wisdom of Galileo's followers. Indeed, we have come to realize that, as someone put it: "If there is a key to the universe, it is not a measure, but a metaphor." All our models are basically metaphors, as I argued. And yet, it seems to me, Forrester wants to reduce the metaphor to the level of a measure. The pernicious grip of empiricism is such that even those who consciously want to depart from its ideology and its consequences inadvertently fall victim to it.

In sum, if adequate descriptions must go beyond the limits of physical descriptions, and if an adequate world model must go beyond the rigid mechanistic box-like modeling, then Forrester has failed on both counts, for he has based his model on physical descriptions and made it operate as a rigid, deterministic, isomorphizing box. Will Forrester ever broaden his model to combat these fundamental limitations? Or will his model remain only a glimmer of a real hope? At the moment, to the skeptic, Forrester appears to be yet another technocrat juggling with formulas and mesmerizing us with the display of computers which, for the present, can do precious little.

6

Systems Epistemology

ALBERT WILSON

THE REQUIREMENT FOR A
NEW EPISTEMOLOGY

THE experience of this century has demonstrated in many ways the obsolescence of our ways of filtering and processing knowledge. We nonetheless tend to hold our methods of knowing as basic, unchangeable, and absolute—in somewhat the same way that two centuries ago we endowed Euclidean geometry with absoluteness—failing to recognize the arbitrariness of some of our epistemological assumptions and values. Specialization and the cellularization of knowledge have generated the requirement for a more comprehensive and integrative approach to our organization of experience, to avoid the body of knowledge growing into some new Tower of Babel. Many of the crises we are encountering in the ecology, in population, in resource use and distribution, in human conflict, etc., are now precipitating the recognition that solutions lie beyond politics and jurisprudence. These crises not only have axiological components rooted in historic religious beliefs but also epistemological components rooted in the current world view of science. Values valid in an age of nomadic migration across the broad plains of an expansive earth—be fruitful and multiply, subdue the earth—are wrong directions for a densely populated finite planet.[1] An epistemology that interprets human experience as an "objective" representation, independent of the experiencer, is not only delusive but tends to avert considerations of the peculiar powers of the experiencer in interacting with the world. Models and simulations of complex systems, up to the world system, show us that there are failures in our comprehensions. Complex systems behave "counterintuitively." Seat-of-the-pants flying does not work for Spaceship Earth. Theobald[2] goes so far as to place the cure for our crises on no less a level than a changed way of perceiving reality. These considerations summarily point toward the timeliness of new value systems, new epistemologies, and a new world view.

The current dominant epistemology is the one associated with science. The precision of definability of this epistemology is not so relevant as its successes in building an extensive and highly reliable fund of knowledge. Though fuzzily formulated, this epistemology has been the most successful of all time. However, within the operations of this success-intoxicated epistemology, there are beginning to be heard some disconcerting signals. The brick-by-brick edifice of scientific knowledge, painstakingly constructed, is developing structural cracks suggesting the need for more comprehensive architectural drawings. New fields of inquiry promise severely to stress science's present frameworks of time, space, form, and substance. ESP or Psi phenomena can no longer be denied or ignored in spite of the difficulties of treating them in accordance with scientific validating and falsifying procedures. The ontological dimensions introduced by psychedelic drugs challenge conventional concepts of "reality" and require a new parameterization of our channels of perception.[3]

As with all epistemologies, the epistemology of science focuses on what it *can* do—which is not always the same as what may be important to do. In the present society, good scientists (i.e., successful scientists) are those who work on problems intuited to have a high probability of being solvable. This strategy is certainly appropriate for a young and incompletely tested epistemology. However, in a well-established epistemology, the displacement of signification-per-importance by signification-per-success imposes biasing restrictions on the directions of inquiry. These restrictions tend to generate a

122

corpus of knowledge that is more likely to map the superficial in the cosmos than the fundamental. The ubiquitous canon, "We *should* do what we *can* do," leads to distortion and imbalance in theory, waste and absurdity in praxis.

Science's obsession with "objectivity" seems both futile and pretentious against the backdrop of its opportunistic approach to signification. "Objective knowledge" is the label pasted on the product of the process that begins with human experience, organizes it into a self-consistent structure, then decants the human experiencer. This de-subjectified knowledge, after being transmitted and stored by human intellects, is applied by human agents to modify the world and its human contents in accord with designs made by human planners. It is not clear why one should seek to remove the subsystem of the experiencer from a world system in order to obtain knowledge of a world system that contains experiencers. It seems rather that the type of knowledge needed for praxis or action must be based on the *total* system in which the action is to be executed. For example, a science of healing that focuses on the human as object to be healed but ignores the properties of the subjective human as healer will find such phenomena as "faith healing" outside its purview. Such a science must either deny these phenomena or term them "miraculous." There may be nothing miraculous about them at all for a science that studies the world system without excluding the properties brought into it by such higher level subsystems as humans.

The epistemology of science has had another unsought side effect. It has robbed man of meaning. In the words of Nobel laureate Alexis Carrel,[4] "Science has made for man a world to which he does not belong." This has been brought about not only through the pursuit of objectivity but through the analytical process of scientific epistemology which is by its nature "a basilisk which kills what it sees and only sees by killing."[5] The atomistic facts that are the excrement of analysis are not the prior-to-analysis holistic system, rich in all of its interior and exterior relationships. We have built a knowledge of the dead pieces devoured and digested by analysis and not a knowledge of the undevoured living world which can never be obtained through this process. Analysis is for the purpose of explanation and explanation is concerned with parts. An explanation is a description of the *contents* of a system and how it works. Meaning, on the other hand, is a matter of relationships, especially relationship to the *context,* arrived at through considerations of the whole. It is not surprising that there is a crisis of meaning in a civilization that is built around an analytic epistemology. It is also not surprising that our models of the world system are concerned only with the inner workings of the system and rarely, if ever, give thought to the system output. What indeed is the output? What is the function of the world system with its life and intelligence with respect to its total context? Such questions are called "unscientific" and perhaps are properly eschewed by science, since they are intractable in its epistemology. But such questions stand nonetheless as primary driving forces for all human inquiry.

One of the most important sources of the requirement for new epistemologies is the need for the capability to validate and significate all types of human experience. The present epistemology of science has proven its worth for experience that is continuous, ubiquitous, and repeatable. It encounters difficulties or an impasse, however, where experience is intermittent, infrequent, or where *ceteris paribus* cannot be invoked. This has resulted in the *quality* of scientific knowledge being dependent on the subject area of the knowledge. The highest quality knowledge under the epistemology of science centers in those disciplines, such as physics, astronomy, etc., where the level of complexity of phenomena is such that repeatability is not obliterated by a profusion of parameters. In general, the quality of knowledge decreases as the system complexity increases, reaching a less than satisfactory state in the highly complex behavioral sciences where unique events that are scientifically untractable may carry the greatest significance. For it is not apodictic that the regular and the universal are sufficient to account for the structure and dynamics of the cosmos and its subsystems. The unique and the exceptional—which for the most part lie beyond the firm grasp of the epistemology of science—may have a significance as great or greater.

The need for epistemologies that will allow us to validate and falsify where samples are small, repeatability not possible, or where unique events override systems parameters, will not necessarily be met through some single all-inclusive epistemology. We should not expect a single epistemology that can equally well subsume sense experience and extrasensory experience; equally well significate mystical experience and practical planning; equally well validate deterministic systems and normative systems. We should seek to develop critical methods for collecting, testing, and signifying, appropriate to each type of system experienced, rather than trying to make one shoe fit all feet and judging the quality of the feet by the fit of the shoe.

One of the central concerns of general systems theory is with methods and frameworks for the unification of knowledge. There can be no unity of knowledge until there are, (a) epistemologies suitable for every type of experience, and (b) a framework—space, time, causality, etc.—of sufficient breadth and depth to permit the formulation of hypotheses and models to account for all the types of experience. A presupposition of systems philosophy is that the world is intelligibly ordered as a whole.[6] Although the world appears to function as a whole, our best representations come out piecemeal. If the world is a whole, there should be some complex, multilevel representation possible. The design of such a multilevel construct depends on a methodology for the valid organization of systems into a suprasystem. Whereas the inverse problem of analytical resolution of a system into subsystems is readily treated by such top-down approaches as deduction, and single-level systems are amenable through induction or statistical procedures, there is no corresponding technique for vertical bottom-up organization. This lacuna is a task for new epistemologies.

Further epistemological requirements are generated by another concern of general systems theory. This is to derive and validate the basic principles and metaprinciples that commonly govern physical, biological, social, ecological, and artificial systems. This task has a resemblance to the epistemological step taken by the Greeks on a more elemental level when they were able to replace such statements as $3^2 + 4^2 = 5^2$ and $5^2 + 12^2 = 13^2$ with the metastatement $a^2 + b^2 = c^2$, valid for all right triangles. But before this could be done the validating process of deductive proof had to be perfected and incorporated into their epistemology. The general systems theorist of today faces a similar epistemological task in the development of suitable canons for validating and falsifying metastatements concerning systems behavior. There are, for example, analogies between linguistic and biological evolution, between the evolution of organisms and of artifacts; there are Zipf's relations[7] between rank and population for cities, or rank and frequency for words in manuscripts, and similar rank-frequency relations in many diverse systems; there is the two-thirds power law relating the sizes of external and internal components of organizations analogous to the surface area and volume of the interior of metric solids.[8] What kind of "$a^2 + b^2 = c^2$" metastatements can be made in these cases, and what level of validity for such metastatements can be established? In other words, is there a general systems theory?

Systems may operate in one or more of three dynamic

modes: deterministic, telic (or normative), and probabilistic. In the past it has been customary to argue which of these three modes exclusively governs the dynamics of the world system. Today we are finding it more useful to postulate the coexistence of all three, and forego the futility of trying to reduce any two to the third. However, various sectors of the intellectual community still prefer to assume the exclusiveness of one mode for their own purposes. Macrophysical scientists tend to assume the deterministic mode applies exclusively in their systems; microphysical scientists, the probabilistic mode; and social scientists, the normative mode. This places the subject area of the bioscientists at the level where modes interface. If biologists opt for an exclusive mode (as most do), they encounter the lacunae of reductionism or those of vitalism. If they opt against exclusiveness, they encounter the epistemological problems of interfaces. In general terms, the modes may be discriminated by some first-order attributes: Deterministic systems are closed, causalistic, reversible, predictable, and receive their inputs on the operational (energy) level. Normative systems are open-ended, finalistic, irreversible, forecastable, and receive their inputs at various control (informational) levels. Probabilistic systems are locally open ended, generally acausalistic, irreversible, unpredictable, and appear to generate their inputs autonomously. (Ensembles of probabilistic systems, on the other hand, are closed, irreversible, and forecastable.) Since general systems theory is concerned with all species of systems, the nature of these modes and their interfaces (or, it must be allowed, their possible reducibility to one another) constitutes a central task for its research.

First, let us take the difficulties with the view of time adopted by science. It is no longer expedient to ignore the finalistic—future influencing the present—aspects of normative systems simply because they cannot be subsumed in the historical notion of time developed in accordance with the causality principle operating in deterministic systems. The biological and social sciences have had to build their models around too narrow a notion of time. Whether or not such difficulties, as are implicit in the "reductionism vis-à-vis vitalism" impasse, could be resolved by a more comprehensive view of time cannot be claimed. But general systems theory should recognize that departures from the "strict constructionalism" in certain frameworks of science—such as time—are necessary if we are to develop the new epistemologies needed for processing and synthesizing all human experience.

Second, we take the matter of values and value systems. Normative systems, in being open ended, are directable through choices made among a set of images of the future. Choices in turn are narrowed by decision algorithms which include in their steps the application of values and value systems. Science prides itself on being value free. This (without the pride) is an overt admission of its inability to cope with normative systems. But this inability derives, as we have seen, as much from the limitations of its notion of time as from science's epistemological value of objectivity. The resulting exclusion of investigations by science into values and value systems has created a critical shortage in our body of knowledge, with derivative malnutritional maladies in our bodies politic.

Related to normative or telic systems is the subject of telos itself. The properties of telos—purposeful or finalistic behavior —have not been adequately investigated. We do not know, for example, the level of complexity at which telos first appears within a system (or whether telos is ever *within* a system but always must bear a contextual relationship). Nor do we know the relation between telos and consciousness or between telos and life. Telos may be an essential concomitant of life appearing on the systems scale at lower levels than consciousness. Or all three may occur in various orders at various levels of the systems scale depending on time and other systems parameters.

Our axiologically and epistemologically rooted crises; the traps of objectivity; the denial or designification of areas of experience that are not amenable to an epistemology designed for the repeatable and the ubiquitous; signification per self-directing successes; the absence of holistic and contextual considerations with the consequent desiccation of meaning; the exclusion of normative systems together with their concomitants of values, value systems, and telos; the need for ways of validating and falsifying the propositions of general systems theory; the need for unitary frameworks of space, time, structure, etc., and for techniques of synthesizing that will permit the unification of knowledge—all these create, individually and jointly, the requirement for new epistemologies and frameworks. This requirement broadens the traditional concept of an epistemology. No longer is epistemological concern limited to what knowledge is and the ways of knowing. It must consider the entire "knowledge system," i.e., the collection, filtering, organization, testing, interpretation, evaluation, recording, and transmission of experience. It must consider the nature of the growth of the corpus of knowledge and the various feedbacks that the existing corpus puts into the growth process. It must consider the morphology of inquiring systems. In all of this, general systems theory not only has basic requirements for new epistemologies and new frameworks, it also has basic contributions to make toward meeting these requirements. The general systems approach appears to provide the best conceptual point of departure for researching the knowledge system. Only a comprehensive, open-minded yet critical view such as that taken by general systems theory will suffice for realizing the epistemological requirements that have been outlined here. The assumptions and aims of general systems theory are facilitating the structuring of suitable epistemologies for many areas of experience and for organizing them into a unitary framework. The close parallel between these epistemological tasks and the aims of general systems theory makes it appropriate to introduce the term "systems epistemology" for this systems-oriented study of the knowledge system.[9] I shall use the term with this meaning in the following sections.

THE CHARACTERIZATION OF EPISTEMOLOGIES

The knowledge system bears the same relation to human society that the genetic code bears to human life. Theories are genotypes, technologies are phenotypes. Innovation takes place in genotypes, testing in phenotypes. The requirement for a new epistemology is thus no less than a call for a genotypic modification, an altering of the knowledge system's genetic code. Genotypic modifications, whether biological or epistemological, are challenges of the highest order. The analogies between the two systems should prove to be mutually helpful to the biogeneticist and the systems philosopher in examining the aims and the consequences of their parallel tasks in "code modification."

We may take a second analogy to further illustrate the systems nature of epistemology. The basic components of an epistemology are a community of experiencers, a set of ways of experiencing, and an aggregate of experiences or things experienced. We may think of the sources of the experiences as transmitters, some of which most experiencers or receivers can tune in, while some are available only to a few receivers at irregular intervals. In this metaphor the various bodily and other senses are the communication channels, and the experiences are the messages received. (It should be pointed out that we deal only with the messages and not with the transmitters. The "true nature" of the transmitters, i.e., the nature of "reality," is an ontological, not an epistemological, question, which is not relevant here.) Knowledge is the organization that the community of experiencers places on the representations of selected subsets of their experiences. An epistemology consists of both the *imposed* and *adopted* rules employed by the community of experiencers for the collection, representation, filtering, organization, evaluation, and application of their experiences. The term *community* implies that the experiencers share, at least in part, the ways of experiencing and, at least in part, the same experiences. This further implies that the members of the community each possess a copy of the *code book* that allows them to communicate with each other the encoded representations of their experiences. The *imposed* rules are the constraints that limit the experiencers in their ways of experiencing and in bringing to consciousness their experiences, i.e., in our metaphor, the basic frequencies and band passes of the channels and the sensitivities of the receivers. The *adopted* rules are the conventions agreed upon by the experiencers for the processing of their experiences. Different epistemologies may be parameterized in part by their adopted rules for validation, signification, etc. These rules, in turn, depend on the relative emphasis placed on certain *epistemological values* such as objectivity, consistency, elegance, etc.

Epistemologies may also be characterized in terms of their "volumes" in three types of space: an experience space, a model (or construct) space, and a cultural space. The dimensions in the experience space correspond to such parameters as the number and properties of the channels through which the experiencer receives his experience (such as the sense channels); the nature of the signals coming over the channels, such as their intensity, frequency of occurrence, duration, and continuity. The properties of the experience space are generally fixed and correspond to the imposed rules governing the episte-

131

mology. However, through the development of sensory-extension instruments such as telescopes, thermocouples, and spark chambers, and through the development of consciousness-extending techniques such as biofeedback displays, psychedelic drugs, and meditative disciplines, the volume in experience space, which is a measure of the experienceable domain of the phenomenological world, may be enlarged.

The model space usually has three dimensions corresponding to the three basic epistemological values of comprehensiveness, precision, and simplicity. The volume in a model space measures the epistemological utility of a model, theory, or explanation.[10] The larger the domain of experience over which the model is valid, the more precisely it maps experience, and the simpler or more economical it is, the higher its overall value. However, there are some trade-offs between these three values. Precision frequently must be bought at the expense of simplicity, and field of view (comprehensiveness) traded for resolving power (precision).

The third space, a cultural or societal space, has to do with the social acceptability of an epistemology. Its dimensions are the length of time the epistemology has been culturally established, the number of people (weighted by their social importance) who subscribe to it, and its success as measured by its ability to meet certain cultural values such as utility. (Successes are also functions of the volumes in model space.)

From these characterizations we see that in both model space and cultural space there are components of the knowledge system that contain values. The knowledge system is thus in part a normative system involving choices that establish these values, a fact contradicting any pretensions to absoluteness for an epistemology. The shape of the corpus of knowledge results from the imprints of these values, giving us the strategy of "value perturbation" as a way to detect unsuspected adopted filters that limit our experience. Different epistemologies not only focus on different regions of experience space but tend to adopt different values for their model and cultural spaces. For example, the epistemology of science and the epistemology that the Greeks called "doxa" and we call common sense, are both primarily concerned with the same experience space—that of the physical senses. (Science, however, is more deeply involved with instrumental extensions of sensory-experience space.) These two epistemologies differ in their model spaces primarily through science's much greater emphasis on precision and less concern with simplicity. The two differ in their cultural spaces primarily through science's emphasis on success and doxa's emphasis on body counts. Only in science and in certain axiomatic theories such as mathematics are there highly formalized validating procedures. Doxa validates through "workability," which as time passes drifts toward validation through tradition or validation through the authority of body counts. The epistemologies used by various "occult" disciplines usually validate directly through the authority of some individual or text. It must be noted, however, that validation by authority is not entirely absent from science. Authority in science, however, operates not on the level of fact validation, but on the level of prescription and proscription of methodology. For example, in the so-called Velikovsky Affair,[11] Velikovsky's facts turned out to be correct but they were opposed because they were obtained by using a methodology unacceptable to science.

Mystical and religious experiences possess no formal epistemologies or validating procedures. The nature of their experiences tends to be highly personal and often much of it is not communicable. Such experience obviously cannot be passed through the filters of repeatability and ubiquity that are imposed rules of epistemologies that are based on the least common denominator of general communicability, as are both science and doxa. The basis for validation in these areas of experience, when it is not some authority, is an "inner-recognition." Inner-recognition is a "gut-level" ultimate in the act of knowing—a sort of resonance with what is true. It underlies the criteria by which we are guided in the construction and testing of our formal epistemologies. It is the court of last and highest appeal, transcending pragmatic criteria, which are always associated with an interval of time in their propositions of validity. It is important, however, to discriminate inner-recognition from the "hunches" and "feelings" and other gestalt perceptions that we lump all together in the English language under the term *intuition*. Inner-recognition and gestalt sense perceptions belong to different levels of intuition. These levels constitute an important sector of study for new epistemologies.

We have noted in the case of doxa the tendency for success to lead to the establishment of the authority of tradition. This is an evolutionary tendency in all epistemologies, perhaps the basic dynamic of the cultural space. But authority on whatever level, once established, diminishes the frequency of appeal to either pragmatic tests or inner-recognition. These important feedback loops in the knowledge system tend to atrophy under the warm glow of past success. An epistemology is one system that cannot afford to be governed by the popular adage, "If you find something that works, stick to it." Vital and effective epistemologies have no orthodoxies; they must be periodically reviewed and renewed on every level.

APPROACHES TO A SYSTEMS EPISTEMOLOGY

How do we begin to meet the requirements for a unifying metaepistemology that will enable us to build a knowledge system containing the essential features of "genetic tapes," and one which, going beyond, provides a suitable "cultural tape"? It is not easy to modify epistemological patterns of thought and practice that have become so ingrained as to be invisible to us. The evolution of these patterns has been slow and painstaking, requiring generations for experiential feedback to effect changes. Now we are asking for a new epistemology to be designed in years, not generations. Such a metarevolution feels subversive on everybody's list. Clearly this is not a task for any one group or school of thought. It can only result from the integration of many ideas and approaches. Four essential steps appear to be involved:

1. Development of awareness of the need for a systems epistemology.
2. Criticizing existing epistemologies and theories to find a fundamental parameterization of the knowledge system.
3. Utilizing this parameterization to generate a morphology of alternative subsystems to function within the knowledge system.
4. Evaluating and selecting suitable subsystems. Integrating these into a systems epistemology.

The first section of this paper contained some remarks applicable to step 1. The second section sketched a few ways of looking at epistemologies relevant to step 2. Since steps 3 and 4 depend on the completion of step 2, we can go no further at this time. The remainder of the paper will discuss a few epistemological miscellanies useful as *Hilfsmittel* in the various steps.

Matters of attitude are among the prerequisites for a systems epistemology. One important attitudinal problem is how to achieve an effective blend of openness and criticalness. Openness is frequently threatening because it might expose work involving a considerable investment of time and effort to inputs that would invalidate it. The response to this threat from openness is often to employ criticism as a wall to shut out innovative inputs rather than as a tool to evaluate them. Proper criticism, however, is based on consciousness of where we are and what we are trying to do and this consciousness does not fear openness, fuzziness, or the tension of deferred validations.

A useful approach that effectively combines openness and criticalness has been described in the rubrics of Zwicky's "Methodology of Morphological Construction,"[12] a methodology useful for syntheses. In Zwicky's technique one employs a temporal pattern of alternating expansion and contraction: an expansive phase of unencumbered imagination of possibilities followed by a contractive phase of critical evaluation and decision among these candidate possibilities. The alternating pattern in time is the essential feature. It is defeating if the imaging and the criticizing phases are not kept scrupulously distinct. Without a season of freedom from criticalness the full powers of the human imagination cannot be released for giving birth to innovations; without a season of focus on criticism, free from the disruptions of novas, no model can be built. Without the temporal pattern of alternating openness and criticalness there could not be the temporal pattern of innovation and construction, innovation and correction on which the growth of the corpus of knowledge depends. Otherwise all would remain either permanently fluid and nebulous or permanently rigid and ossified.

The ability to employ such an alternating pattern depends on an attitude that can withstand the tensions of postponed resolution of antithetical concepts (admittedly a difficult stance for the "now generation"). Resolution and decision are required for praxis, not for theory. Action and implementation demand the convergence of option space; but it is otherwise profitable to keep the stock of possible alternatives as rich as possible for as long as possible. One of the longest unresolved tensions in the history of science had one of the most fruitful resolutions, when finally it came. This was the particle-wave tension and its subsequent resolution through quantum mechanics. Had not Huygens's wave model possessed such a broad experiential base, it is possible that certain of Newton's followers using their customary Cromwellian clout would have succeeded in resolving the particle-wave question in the seventeenth century in the usual manner, through repression. However, the cosurvival of the two antithetical viewpoints provided a stimulating and fruitful tension within physics that delayed resolution until it could be made through synthesizing rather than through opting. Alternative models and perspectives are useful even when their claims for adoption are not so nearly equal as in the wave-particle case. Alternatives often provide us with stereo vision.

Postponed resolution of epistemic tensions would have an important effect on the manner of growth of the corpus of knowledge. The present manner of knowledge-growth resembles that of crystal-growth. Both grow through a process of epitactic accretion to the outer surfaces of the existing bodies. In epistemology, explanation of the new is always in reference to the terra cognita of the well-established corpus. In fact, "to explain" generally means to relate to the familiar. The custom of insisting on this one restrictive type of relation—linking new discovery to the main corpus—results in the restriction of growth to epitaxis on a single continent of knowledge. In this process the "islands of knowledge" that cannot immediately be related to the main body have small chance of survival. Only when an island provides some compelling utility or economy can it survive without being explained. For example, Heaviside's operational calculus was too useful to discard even though it could not immediately be validated. The Titius-Bode law of planetary distances has survived over a century without explanation because it discloses an intriguing simplicity of organization. But the general rule for new experience is: "Be explained or perish." If the tension of unexplainable islands could be sustained then epistemic growth could proceed through the growth of each island and whenever possible through the relating of islands to one another without the necessity of their being related to the continental corpus, i.e., of being explained. A current example of an island of knowledge is the UFO phenomenon.[13] The non-epitactic approach to UFO's would be to postpone explanation in terms of psychology, extraterrestrials, or whatever, and synthesize the various patterns contained in the observations; then utilize the patterns to provide the specifications for the design of a "flying saucer," going as far as is possible by employing known relations, and in this way isolating the lacunae in our knowledge. These lacunae will probably provide the keys for a future explanation. But since UFO's cannot now be explained, the epitactic process chooses either to dismiss or suppress the subject instead of encouraging the island to grow. In this case trouble was even taken to establish a hierarchy of committees to validate the suppression.

The basic question regarding islands is not *explanation*; it is *authentication*. To authenticate a body of experience usually means to establish the existence of a nonillusory, nonchance, internally consistent set of events. In a systems epistemology that is capable of dealing both with illusion and chance, authentication is better defined in terms of the existence of some critical size for relational patterns, whether or not illusion and chance be present. The epitactic approach, in focusing on the features that relate new experience to the main body of knowledge, gives a preferential status for purposes of explanation to those systems that, for whatever historical reason, happen to have been examined first. Since the first systems to be successfully studied scientifically were those lowest on the systems scale—physical and chemical—explanation for new experience must be made in terms of these systems. Thus reductionism is an imperative of an epitactic epistemology. If other systems than chemistry and physics had

had this primacy of study they would also have had primacy for a role in explanation.

When Apollo 8 brought back the first pictures of the blue globe of Earth floating in space, we received a new paradigm for our epistemologies. Instead of viewing structures as being based and dependent on some main body that is foundational for all components, we now can see that a foundation is but one more synapse in the structure, and like all the other links and synapses, it too floats. Relational links of every sort between synaptic islands are paraexplanations. Our epistemic structures will be richer and more comprehensive insofar as we allow the great variety of linkages that may exist between various islands to enter, whether or not these linkages exist between each island and the primary corpus. This is, in the language of systems commonalities, the basic aim of general systems theory.

In summary, the requirement for new epistemologies is primarily to *supplement* the epistemology of science. The past successes of science have encouraged us to endow it with the future promise of unlimited success in solving all problems and leading us to the realization of whatever goals we seek. But this is unfair to science. Those working closely in and with science do not make such claims nor encourage such expectations. In fact, the more closely one works with the epistemology of science the more clearly one sees its limitations— limitations of the sort pointed out in the first section of the present paper. However, the call for new and supplementary epistemologies is not likely to be heeded in face of the myriad successes of science. But success does not get corrected, and we may expect that the destiny of science is to experience the "failure of too much success." Before this happens, those concerned with preserving whatever of positive value has been achieved in the cultural tape must begin to make the needed corrections and to broaden the base for the critical acquisition and evaluation of knowledge of whatever nature; new episte-mologies, one appropriate for each domain of inquiry, must be structured; and the whole unified under a comprehensive framework that permits experience of every sort to be modeled. This set of new epistemologies, together with that of science, and the coordinating framework for their synthesis, is what we seek here under the designation "systems epistemology."

7

United Studies

HÅKAN TÖRNEBOHM

How does this paper relate to the main topic of this volume? It is a paper about how to study inquiring systems and how one may operate in such systems, bringing about fusions if that is desirable. Complex systems with human components will change very much if inquiring systems are appended to them. Inquiring systems produce innovations which give rise to sequences of innovations. We may run into great difficulties if we do not understand innovative systems. World system research is one example of an innovative system appended to complex systems with human components. To map the research itself is therefore of great importance.

INQUIRING SYSTEMS

Studies about a territory X that is a part of the real world, may be described as follows:

A group of active researchers carries out three kinds of activities. They make investigations in which they draw maps of X. They report on forthcoming results to each other and to outsiders. The tasks of investigation and reporting are governed by steering activities. Forthcoming results are submitted to a critical review in which criteria are employed based on the ideals of science to which the researchers adhere.

An inquiring system has also a recruiting department in which new researchers are educated and trained.

Figure 1 pictures studies of a territory X.

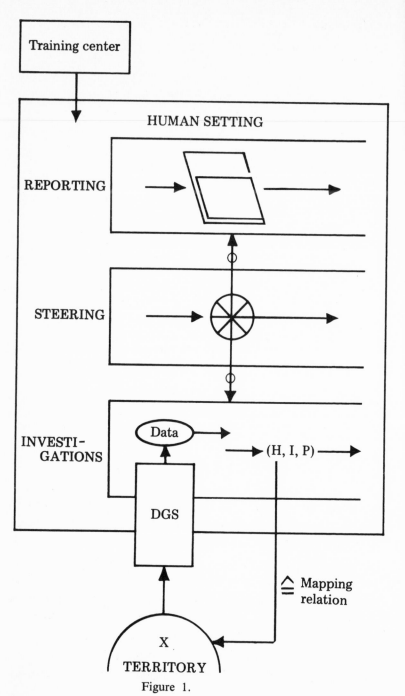

Figure 1.

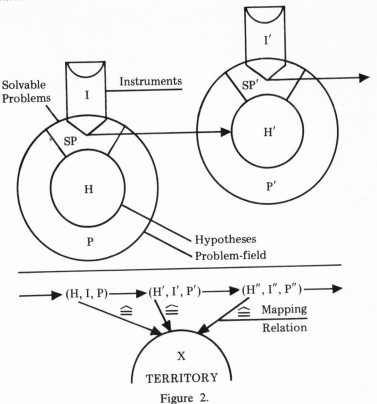

Figure 2.

INVESTIGATIONS

Investigations may be described as sequences of transformations in which complexes of informations about X, *h*ypotheses H, a field of *p*roblems P surrounding H, and *i*nstruments for solving them are transformed into other (H, I, P) complexes. The body of information H comes from reports from earlier investigations and also from data-generating systems through which messages from the territory pass into stores of data.

The instruments are either hardware tools—instruments of measurement and laboratory appliances of other kinds—or they are software tools—techniques which have been designed and constructed by mathematicians, statisticians, and others. Figure 2 pictures one move in an (H, I, P,) sequence.

H contains, among other items, data and hypotheses originally framed as tentative answers to questions about X. The hypotheses may be divided into different epistemic classes as follows:

E 1. contains hypotheses for which no testing procedure has yet been designed.

E 2. contains hypotheses for which a realistic testing plan has been designed but not yet been implemented.

E 3. contains hypotheses which have been tested, but which have not yet been accepted by the researchers working on them.

E 4. contains hypotheses which are regarded as credible by those working on them but which have not yet been accepted by the community of experts in the field.

E 5. contains hypotheses which have been accepted as public knowledge by experts.

Hypotheses belonging to any one of the classes E 1. to E 4. are problematic. The problems associated with them are different in nature, depending on the epistemic class to which a hypothesis H belongs.

If a hypothesis H belongs to the class E 1., the corollary of H is a design problem. A hypothesis H in E 2. is associated with the problem of implementing a testing plan.

The outcome of a testing procedure may be that a hypothesis H is rejected wholesale, or that H is refined into a hypothesis H' in which a good deal of the information carried by H is preserved, or that H is judged to be credible by the researchers working on it.

The problem associated with a hypothesis H in the class E 4. is to get H accepted by a court of expert judges. The task is to argue in a convincing manner in favor of H. A description of the testing procedures is an essential part of such an argument.

146

If a researcher fails to convince competent experts to accept H, the hypothesis moves down to a lower epistemic class. When a new hypothesis in one or the other of these classes enters into a body H of information about X, H changes into H'. At the same time, the problem-field P is transformed into P'. New information entering H' gives rise to new problems, which may give rise to changes in the store of instruments. New instruments are imported from outside or they are designed and constructed at home. Old instruments may be refined into more effective ones. The sector of soluble problems is thereby affected.

STEERING

Perspectives. One factor which has a major influence on studies about a territory X is the ways of viewing X.

A territory Y has already been mapped. A territory X has not yet been investigated. The researchers look at X as being similar to Y. That is their perspective on X, which makes it plausible for them to believe that maps to be drawn of X will be similar to maps which have already been drawn of Y. A world picture determines in which directions researchers are likely to look for positive analogies.

This strategy, if successful, widens the scope of knowledge which originally maps a limited region of the world. In that way general knowledge will be acquired, say, about inter-actions, about processes, about compositions of systems, about regulations in living organisms, etc., which will be incor-porated into many pages of a world map.

To see profound similarities is a great skill. Newton saw that the motions of planets, and of apples when they fall from trees, are similar. He conceived of them as processes in sys-

STEERING FACTORS IN THE HUMAN SETTING
AND THEIR SELF-PERMANENTING BY THEIR
INFLUENCING THE TRAINING CENTER ACTIVITIES

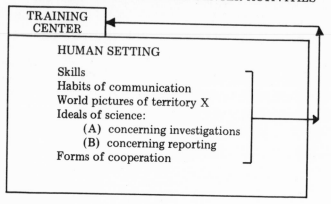

Figure 3.

tems in which the components interact in accordance with a universal law of gravitation.

Perspectives determine the process of seeking information from previous research: they determine what problems are to be regarded as important. They play a major role in the process of concept and theory formation.

Ideals of science. Ideals of science, like world pictures, are acquired during the training process.

Prevalent ideals become relatively permanent. There are general ideals of science which last over many generations in a field of research, such as the ideal of a mathematical and experimental science.

There are also more special ideals. The personal examples furnished by admired leaders in a branch of studies act as paragons of good research. Certain written works are regarded as masterpieces; they set standards for good reporting. As a result a large portion of the research carried out in any field has an imitative character.

INTERNAL CRITICISM

Forthcoming results of an investigation, as well as reports, are submitted to critical assessments in accordance with criteria which are based on operative ideals of science. The internal criticism of the forthcoming results of investigation has two main functions:

(1) to spot errors made in the innovative phases of the investigations;

(2) to legitimize or authorize hypotheses which thereby acquire the status of full-fledged public knowledge. The authorization of cognitive items includes indirectly an authorization of research procedures, described in the same report in which an investigator advances a new piece of information. Corresponding to these two functions of internal criticism we may ascribe to critics the role of problem generators, of finding flaws giving rise to problems of amendment, and, to some critics, the role of umpires. Authorized umpires embody the operative ideals of science in a field of study.

DIAGNOSIS OF TROUBLES

It happens now and again that researchers find themselves in situations which are quite different from the normal situations in which they encounter soluble problems.

Here are some examples:

1. Anomalous exceptions

A razor blade does not sink when it is put gently on the surface of water. Yet it ought to sink according to hydrostatics. The phenomenon is anomalous from the point of view of an established theory. What is required here is a new research

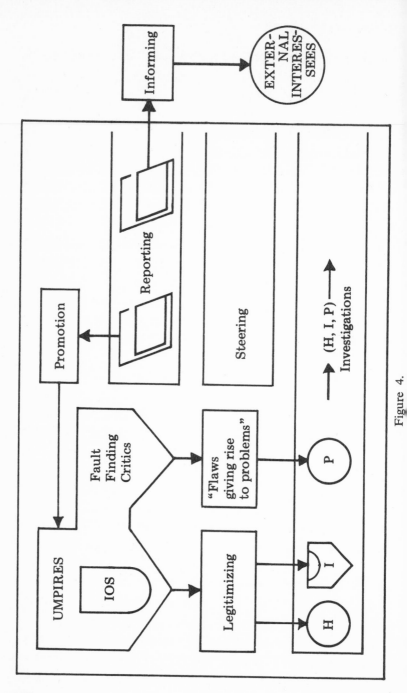

Figure 4.

program concerning surfaces of liquids. This anomaly disappeared when knowledge about capillarity had been worked out.

2. Internal and external clashes

The Newtonian theory of space and time is well adapted to the remaining parts of Newtonian dynamics but fits less well to the remaining parts of electrodynamics. Yet the Newtonian theory about space and time forms part of classical electrodynamics. The Michelson-Morley experiment led to the discovery of an internal clash in classical electrodynamics. This flaw was repaired by Einstein, who invented a new theory of space and time and combined it with the remaining parts of classical electrodynamics, translated into a new formalism. The result was relativistic electrodynamics. This new theory clashes with the old Newtonian theory of gravitation. A new reform program was called for, designed, and implemented. Its output is the general theory of relativity.

3. Residuals

A vast collection of empirical knowledge has been brought together. In spite of strenuous efforts, theoreticians have failed to assimilate this knowledge into organized bodies of theories. An example is the mass of knowledge about spectra which had been collected by generations of scientists before Bohr invented a theory into which part of this knowledge could be assimilated.

This list of troubles is a partial one; it serves mainly to clarify what I mean by "troubles." Troubles call for diagnostic and therapeutic efforts. The therapeutic efforts may give rise to new research programs which may lead to extraordinary innovations. Particles of sand are very irritating to mussels— yet pearls are made from them. Some troubles are like sand in mussels.

151

Some troubles are branching points in (H, I, P) sequences. Perspectives on old territories change, new territories are explored. Families of theories of relativity are outputs of reform programs. New families are constructed in which hitherto unexplored territories are mapped. Quantum mechanics is such a family.

REVIEWING BEFORE REPORTING

Research situations are often opaque in foresight, but may become transparent in hindsight. One learns from mistakes and lucky strikes. Reporting to recognized umpires serves the purpose of getting one's results authorized. A truthful description of the actual route which led to a result may actually be a poor argument on its behalf. A better argument is a rational reconstruction of the route, from problems to their solutions. Reviewing in hindsight plays, therefore, a part in the process of reporting and is likely to lead to improvements in skills and competence.

The steering activities as well as steering factors are pictured in figure 5.

THEORY AND EXPERIENCE

In this section I shall consider an important kind of united studies, *viz.*, linkages between empirical and theoretical work. Isolated pieces of knowledge have less value according to prevalent ideals of science in mature fields of research than pieces of knowledge assimilated into organized bodies. In order to enhance the value of a piece of knowledge it is desirable to build it into an already existent system of knowledge or else

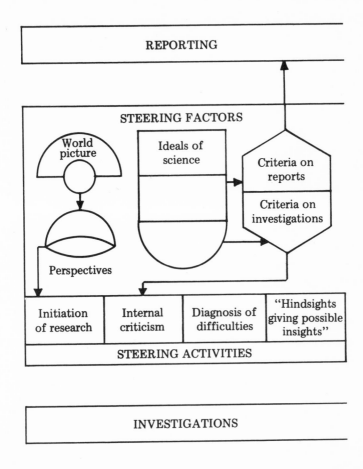

Figure 5.

to invent a system of hypotheses in which this piece of knowl-edge, together with other pieces, can be accommodated.

Work towards these ends constitutes aspects of theoretical science; empirical investigators are concerned with other tasks.

They design and build data-generating systems from which data of a high epistemic quality about a territory X may be obtained. They have their own exploratory research programs aiming at the establishment of pieces of knowledge about a territory. Experimentalists and others engaged in empirical investigations have their own perspectives and ideals of science. They are concerned with the following qustions: How should effective testing plans for specific hypotheses be designed? How can data about this territory be obtained? What kind of data should we try to obtain? The answers to these questions depend on the context—for instance, a testing plan for a hypothesis, into which the data will be fitted. How can data be controlled? How can systematic errors be eliminated?

Reports about hypotheses of the highest epistemic class go from E-centers to T-centers, where they give rise to theoretical problems. Theoreticians try to assimilate them into already existing organized bodies of knowledge.

If the theoreticians adhere to a deductivistic ideal of science, according to which explanatory patterns should be deductive arguments and theories should have the form of deductive systems, then an assimilation of a piece of knowledge has succeeded if a piece of information k' is deduced from premises of the highest epistemic class, where k' has the same content as k but possibly another form. A deduction of k' from full-fledged knowledge is an explanation of what k' and also k asserts about the territory under investigation.

If k is new, it is not possible to find a set of premises from any already established body of knowledge from which a piece of information with the same content as k can be deduced. Theoreticians must therefore invent one or more supplementary hypotheses to be added to items selected from old knowledge. These hypotheses start their existence in the lowest epistemic class. Deductive patterns in which they are contained are therefore not full-fledged explanations to begin with. In order for an epistemic criterion of adequacy concerning explanations to be satisfied, it is required that hypotheses, framed by theoreticians engaged in a process of assimilation, should be tested. Theoreticians have no direct access to data. They must therefore call on assistance from E-centers.

It often happens that the hypotheses framed by theoreticians cannot be tested in a direct manner. It was not possible to submit Newton's hypothesis of gravitation to a direct test, framed as it was in an attempt to assimilate Kepler's laws of planetary motions, nor was it possible to do so with Mendel's gene hypotheses or with Planck's quantum hypothesis or with Einstein's photon hypothesis. All of these hypotheses were invented in attempts to assimilate empirical laws.

What happens in such situations is that the theoretician asks the following questions: Given that these established pieces of knowledge and my new hypotheses are true about such and such objects, what else must be true about them? Answers to such questions are obtained by means of deductions resulting in one or more hypotheses. These deduced hypotheses may belong to a class for which it is possible to design effective testing plans within an E-center.

Now the E-people carry the ball. They design testing plans and carry them out. They arrive at their own verdicts and try to make their umpires share their convictions. Suppose that the deduced hypotheses move up to the highest epistemic class —then they provide favorable evidence for the recalcitrant hypotheses. If many such deduced hypotheses are corroborated, the umpires of the theoreticians will authorize deep hypotheses, such as Newton's hypothesis of gravitation, that are framed by theoreticians in the process of assimilation. When that has happened, new knowledge has been created fitting neatly into a deductive system.

What has been said could be illustrated by the well-known story about the development of Newton's theory of gravitation.

In the process of framing deep hypotheses, theoreticians are guided not only by formal clues, but also by their general world picture. What kind of system are we now dealing with? What kind of mechanisms could produce those phenomena which our laboratory friends have discovered? Are the interactions mediated by transports of some sort? If so, what is transported and what are the laws of its transport? Theoreticians look with mental eyes into black boxes and imagine things about what may be going on. Then deep hypotheses may flash up in their minds.

Theoreticians work not only on assimilation problems: they also have their own research programs, aiming at the construction of coherent well-structured families of theories, in which theories about space, time, matter, motion, causality, regulations, or teleological structures, are built in. In the implementation of these programs they give the laboratory people tasks to carry out; in addition, they offer them software instruments in the form of theories which can be exploited in constructions of hardware tools and in making experimental designs. They also help them in mapping their instruments. Maps of data-generating systems are absolutely essential if data is to make sense and if it is to be controlled. Data are theory-laden.

STUDIES OF STUDIES

Studies about a territory X may be called a system $\widehat{X}$. $\widehat{X}$ is the territory for an inquiring system; the latter I will call a system $\widehat{\widehat{X}}$.

Studies about studies of X can be described in the same categories as any other inquiring system. There is a human setting, a training department, and a territory. There are investigating, reporting, and steering activities. There are ideals

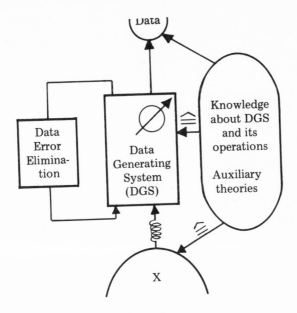

Figure 6.

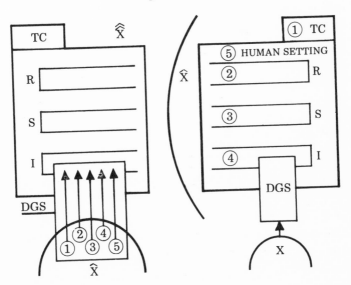

Figure 7.

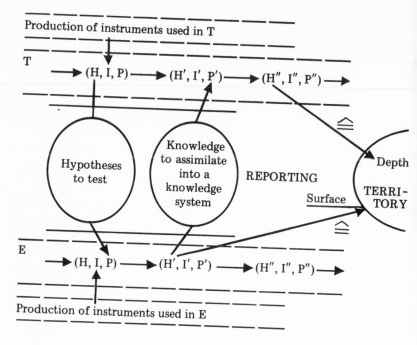

Figure 8.

of research and of reporting. There are perspectives in which various facets of inquiring systems are put into focus. There is also a data-generating system belonging to $\widehat{\widehat{X}}$ through which information about $\widehat{X}$ enters into the activity of investigation in $\widehat{\widehat{X}}$. Various facets of $\widehat{\widehat{X}}$ will now be considered.

DATA GENERATION

To obtain information about $\widehat{X}$, we need to obtain trustworthy information about various components of that system. To obtain such information we need five positions of observation where informants thoroughly familiar with the inquiring system $\widehat{X}$ are placed.

158

These positions are as follows:

1. The training center. The informants take part in the activities going on there in the roles of teachers or students.
2. The department of reporting. The informants study relevant reports and examine the procedures of reporting.
3. The department of steering. The informants take part in the steering activities.
4. The department of investigation. The informants are participating observers.
5. The human setting. The informants make field studies on the X-ologists and report on their habits of communication, their world picture, their ideals of science, their forms of cooperation, etc.

Information must have a purpose. The information about $\hat{X}$ coming from observers in the above five positions of observation should contribute to the aims of investigation in system $\hat{\hat{X}}$. It is therefore important that informants placed in $\hat{X}$ take part in the activities going on in the system $\hat{X}$. Good informants should be at home both in the system $\hat{\hat{X}}$ and in its territory.

STEERING FACTORS

Perspectives. Previous sections of this paper are essentially sketches of perspectives on inquiring systems, including X-systems.

Ideals of science. We conceive of two goals for studies of studies. One goal is to improve the self-images of inquiring systems, including those in which other such systems are studied. Self-images are improved by maps of inquiring systems in which their major features are portrayed with a high degree of fidelity.

Another goal is to assist in critical assessments of the performances of inquiring systems.

Combining the two goals we may say that studies of studies should constitute inquiring systems, with their critical functions directed to targets in their territory. The critical activities should be linked to critical activities in the inquiring systems themselves, including their targets.

INTEGRATIVE EFFECTS OF STUDIES OF STUDIES

Studies of studies may have beneficial effects on activities going on in inquiring systems. It may happen that researchers in one field are unaware of important developments in another field, which by their own criteria can be beneficial to their work.

In may happen that a temporary or permanent fusion of one or several inquiring systems may bring forth results which can be expected to be valuable to several groups of interessees.

In order for studies of studies to have effects such as these, it is evident that they have to be united. Studies of several $\hat{X}$-systems $\hat{X}_1 - - - \hat{X}_k - - -$ are combined into studies of a complex composed of $\hat{X}_1 - - - \hat{X}_k - - -$. In this way connections between them may be discovered which were not known before; it may also be found that establishing clusters of such studies might prove to be beneficial.

There are several kinds of studies of studies in existence: e.g., philosophy of science, history of science, sociology of research and research establishments, psychology of research, etc. People trained in these fields ought to combine their skills and efforts in unified studies, in which more aspects of inquiring systems are examined than any one participant can study on his own. If experts on studies of studies combine their efforts into joint research, and if interested persons become convinced that this is desirable and feasible, it may come about that existing studies of studies are integrated into united studies.

160

8

*The World System
and Human Values*

RALPH WENDELL BURHOE

THIS title, "The World System and Human Values," might be translated into more conventional forms of our language to give it richer meaning, such as "a model of the way the world works and what this implies about human fate," or "a model of the ultimate powers with implications for human hopes and duties," or, in traditional Western religious language, "God's nature and man's salvation."

I shall introduce the relationship between system dynamics and traditional theology by the end of this paper. But in order to provide some initial orientation, I suggest that systems analysis is essentially talking about the same reality that men have called God, especially when our analysis rises to higher or more comprehensive systems in our hierarchy of systems to include the world system or ultimately the cosmic ecosystem. That is, when we are talking about the system that ultimately determines human destiny, we are talking about God.

Further, I join Jay Forrester in suggesting that there is a wisdom and function in our traditional religions for maintaining man's long-term values, which we have partly lost, but which should now be retrieved to save us from the disastrous outcomes predicted by systems analysis for the next century in terms of overpopulation and overconsumption with depletion of resources and consequent decay or disruption of human life, individually and socially.[1] To be effective, of course, the religious wisdom must be translated into models or symbols that are acceptable and meaningful currently. I shall be seeking such a translation before I finish.

163

WORLD SYSTEM

First, to make clear what I mean by "world system." The term "system" refers to a collection of parts that form a whole. I want to make something of the distinction commonly made between a *system itself* as the "real world" and a *model* that supposedly represents the "real world."

As a model of the real world, the term "system" refers to a system of concepts, pictures, symbols, words, or objects, and rules for operating them in a time sequence. With such models people play games in their heads, on game boards, or in computers, with outcomes which supposedly parallel and hence potentially predict states of the real world.

The more seriously a person takes his model, the more it seems to him to be a "real" or "objective" system, the true nature of the thing or the world. What I want to suggest, in line with much philosophy of science, is that we never get away from models in our head and that "reality" is never known for itself. The field of my awareness, especially when unfamiliar patterns bombard my sensory system, may be a blooming, buzzing confusion of innumerable and unrelated sensations for which I cannot account in any rational or "systematic" way. I simply grant that these are data, the givens, the phenomena of my experience. When I begin to conceive some order, system, or relatively invariant patterns of correlation among these phenomena, then I say that I have discovered the underlying reality. I suggest that the phenomena of experience which we designate as "real" are thus differentiated as we find multiple paths for correlating certain elements of experience to fit with minimum deviations certain corresponding invariances of symbol systems or models in our heads. I join P. W. Bridgman in his insight that we never get outside our own realm of experience or awareness, and the route to the "objective" or "real" is simply to find more than one route to confirm one's hypothetical model of the way things are.[2] These routes include, of course, my sensations or awareness of what I symbolize or call other people.

However impossible it may be to get out of our own experience, that experience itself has provided us with a rich way of talking about the world as an objective reality, or a system that remains relatively constant in the midst of the ever-shifting flux of the data of our primitive awareness of a blooming, buzzing confusion. We may observe that a baby in his crib finds correlations among the various patterns that a heavy, hard, red ball makes upon successive retinal, tactual, and kinesthetic exposures that are somehow miraculously assembled in its central nervous system, and produce the conviction in the "mind" of the infant that he is dealing with a reality which he later learns from his mother or brother is an object— "a heavy, hard, red ball" that is not in his head. But whether I experience or talk about my own experiential field of awareness, or about the baby's awareness, or about the ball as an object in itself, it is in reality all the time only I who is experiencing and speaking. We never get away from our field of experience, even when the "objective" models or systems, which we project upon the buzzing confusion, provide us with some degree of invariance, order, or system that is shared in our discourse with others.

My own view of the world system goes further than this recognition of the unity of "subjective" and "objective" to note what I have learned from reading in evolutionary theory. I may properly refer to my genotype, as I do to my brain patterns or to my conscious field of awareness, as a record or model of the environment, produced by environmental selection of feedback, such that, in any future interaction with that environment, it provides the "information" that maintains or replicates a viable organism of the same general pattern as before.

The genetic information makes possible the existence of a relatively invariant phenotype in the midst of an environment that is a buzzing confusion. The genotype is the environment's model of a viable organism, or a relatively stable open system. The genotypic information encoded in the DNA is, so to speak, a mirror reflecting or remembering certain relatively stable potentialities in the world system around it. Life and man are made in the image of the total world system, and are partial reflections of it.[3]

Furthermore, the information transmitted from my culture to shape my brain patterns is also a partial model or image of the world system. It too provides me with information that allows me to maintain some degree of invariance, that is, to respond to the variations in that world system without losing myself. Thus the input of stabilizing frames of reference into the cybernetic norms of my central nervous system from my genotype and my culturetype provides a memory store or image of the world system that is a model of the world relative to me. It is a model such that it is statistically probable that my behavior in accord with the norms of this model will make me and my species viable forms, and will also incidentally provide me with an internal or "subjective" model of myself and my world that is so invariant that I call it "objective."

By these few paragraphs I am trying to establish the view that there is a correlation between the symbol systems (models or information) inside of living creatures and the "real" environment which surrounds them, a correspondence that in the case of the genetic information has been selected for a few billion years, and is pretty good. Another way of putting it would be to say that man is a partial model of the world, created by the world itself. The objective and the subjective are ultimately the same thing. A world system is a model of the world and a world system is also the relative invariances of the world itself as its behavioral variations are filtered through a model of itself structured by the memory of a long experience of it by that part of it which we call man.

It should be noted that men not only have machinery inside their cells and heads that provides a model or reflection of the world as a system, but men can also make machinery outside their heads to provide a model or system of the world. Simple games with sticks and stones and dice and cards and coins are models of aspects of a world by which men can move the pieces according to rules and win or lose. Languages are models of the world, or systems of the world. Science is a special form of language, or symbolic game, which provides a cultural extension of the untutored brain so as to enable the brain more adequately to reflect certain invariants or systems of order in the blooming, buzzing confusion of the world. When the symbolic logic of the sciences is fed into a computer, the computer can operate according to the particular grammar or logic built into it, and the computer can then provide a model of what is going on in the world. Jay Forrester and his colleagues at the MIT System Dynamics Group recognize that the model of the world built into the hardware or fed into the software of computers is, like a genotype or a language, only a heuristic device to help provide some degree of reasonably simple system or order by which to comprehend or compute a future condition of the system. I will venture to suggest that computer projections, like primitive prophetic projections and still more primitive random variations of DNA chains, are trials whose errors will be weeded out by a process of natural selection to fit the "real world" and eventually be reduced to viable, useful, continuing, or "objective" images of the world system.[4] Computers also evolve, and we can speak of their evolution into more complex systems as essentially analogous to the evolution of biological systems: the information within the system increases to adapt to (maintain a relatively stable open system in) a wider range of the environmental circumstances (the "real world").

What is important to emphasize is that today the sciences have become cultural evolution's best yet or most rapid way of acquiring valid or viable new information or models about man and his world, possibly as fast as doubling the amount of accumulated information in every decade or two. Perhaps, apart from its rapidity, it is no better than the successively slower accumulations of information in cultural evolution operating prior to modern science, developed a few centuries ago, prior to written language, developed a few thousand years ago, prior to vocal languages, developed some tens of thousands of years ago, prior to DNA codes of information, developed some hundreds of millions of years ago.

But the *rapidity* is important, for what we have not yet accomplished is a matching of this rapidity with which scientific understanding allows us to develop new technologies for manipulating ourselves and our world by an equal rapidity for evolving our long-range values or wisdom. It is this mismatch which is producing the horror-scopes of the Club of Rome, of *Nineteen Eighty-Four*, of *Brave New World, Counter-Culture,* and *Where the Wasteland Ends,*[5] as well as the increasing anomie of sensitive and intelligent young people. It is on this account of the mismatch and the consequent threat of disastrous breakdown of men (both as individuals and societies) that we need to examine more carefully the question of human values.

Before I discuss the nature of human values, I must say one more thing about systems, and in particular about the *value* of a system which is in a man's head, as a model of his "real world." The model of the world in man's central nervous system (whether the model was shaped by the pattern stored in the genotype, or stored in a culture, or by the most recent, privately learned patterns) is like the organic self, and the world, a dynamic, relatively stable, open system. We call it a cybernetic and homeostatic system to suggest the nature of a machine that is able to maintain a steady state or dynamic equilibrium, while all around it is changing and losing its order, organization, and life. This means that the negative feedback to inform the central control mechanisms about departures from the established norms in the brain operates to inform the individual what needs to be done to return the system to the homeostatic norms or operates more directly to motivate unconsciously the behavior necessary to correct the departures.

This model or picture we have of the world system is an ordered complex system of relatively invariant or stable patterns like the organic self in this respect. If we put on spectacles that turn the world upside down, we nevertheless soon learn to see it rightside up. A child in the United States who is informed that the world is ball-shaped may worry about the Australians falling off if they don't cling to the trees with their legs hanging "down" towards the sky, until he matches this information with information that the direction of the force of gravity is not established with respect to a point in the stars but with respect to the center of the earth. The brain, as J. Z. Young has pointed out, is a magnificent homeostat.[6]

The supreme criterion for the value of any conceptual system, any model of the world, is how adequately it serves the living system to adapt to the requirements of its world. Its dynamic projections of dangers and opportunities are what keep living systems in being, that is, in their relatively stable or steady states. The information in the model or conceptual world system must then reflect both the "reality" of the environment and the needs and opportunities of man within that environment. Values for a living system, then, are a part of knowledge or accumulated information. The great error of the centuries since Kant and Hume is the divorce between science and values, or facts and values, a divorce which if kept up much longer will bring us to our end. Values are a special class of fact or information that give the knowing or living agency the capacity to remain in being.

HUMAN VALUES

Human values, then, are a part of our world system, whether we are talking about our world system as a "symbolic model" in our heads or as the "real thing." As we have seen, the symbolic model is a part of the "real world." That is, human values are not what the philosophers call "sui generis," or a part of another world that is different from or independent of the "real world." Human values are the cybernetic norms or goals of human behavior. Good values are those norms that keep the living system in being or growing. One can locate these norms in the "real world" of the scientist. Our scientific models of man now show the central nervous system as the locus where behavior is determined and where norms are set and behavior correspondingly controlled. These models show that human behavior (including the feelings of hope and pleasure, of fear and pain, and all esthesias and intellections) is a product of dynamic neural patterns. The neural patterns are shaped by three major levels of information or norms: (1) genetic information, mixed with (2) socially transmitted cultural information, as these two interact within the (3) boundary conditions set by nonhuman elements of nature, which also constitute a body of information or norms that shape living systems. All three of these systems are patterned segments of energy flows in the cosmos. Collectively they provide the information (the form, norm, or pattern) of human life.[7]

The wisdom of these human norms or values—from both

our genetic and our cultural heritage—has thus far provided our emergence from one level of complexity to another within the energy-flow patterns of the earth during the past few billion years. It is a part of the picture that a tremendous body of information that provides the wisdom for life—millions of times more than that of which any man is capable of being conscious or aware—is cumulated in our heritage, which has established the values or norms of the cybernetic or homeostatic hierarchy of mechanisms that we call man. One of the delights of modern science is its revelation of the marvelous cybernetic machinery or information that does indeed endow us with dynamic homeostasis[8] or life, progressing to ever higher levels of information (or negentropy relative to the environment) in a world where entropy is always increasing.

It is useful to divide human values into two categories: (1) the already established or existing norms or goals that have directed our behavior and provided us with our life up until now; and (2) the potential or virtual norms that are yet to be established and which our projections of future conditions suggest will have to become embodied if we are to survive or to flourish under certain future conditions. Living systems are homeostatic, but homeostasis is a dynamic process with ever-changing patterns, and the homeostatic norms or values themselves evolve. That is, the genotypes of Homo have evolved from the genotypes of the most primitive living patterns. Hence, values are constantly changing. And in human cultures we find a similar evolution. For instance, the structure or pattern of language evolves, even though it maintains a continuity (metastable or relatively stable state) which defines a living system as opposed to a chaos or void. It should be noted that the informed order or structure which shapes living systems is a complex hierarchy of levels where each new level requires as its foundation the whole complex system of previously established levels. This means that man

171

requires plants, and plants require carbon molecules. But throughout the range, evolution makes no radical jumps; it takes one small step at a time. The norms or values or goals that inform living systems can at the same time be said to be a billion years old and yet also in their newer adaptations they are just being born. This is a picture of evolving life as the process is revealed by the sciences.

In a general systems analysis, one may view living systems as a class of systems among other systems. During the past two or three decades an enormous amount of new information has been consolidated concerning the relation of living to non-living systems. Among the landmark contributors to the new views are A. I. Oparin, whose *Origin of Life* (1938) broke new ground for an understanding of the continuity between chemical or molecular and biological evolution. Erwin Schrödinger pioneered our understanding of living systems as open systems structured by molecular memories, evolving in a direction contrary to the trends of nonliving, closed systems under the second law of thermodynamics.[9] Norbert Wiener provided more on the "negentropic" character of living systems and on how they maintain their complex structures by means of cybernetic mechanisms.[10] James D. Watson and Francis Crick pioneered in unraveling the molecular character of the genetic code.[11]

In my opinion, one of the classic contributions to understanding both the nature of evolution and the selective processes, as well as the common basis of life systems and nonliving systems, is J. Bronowski's "New Concepts in the Evolution of Complexity."[12] He breaks the Gordian knot of the apparent contradiction between the trend of evolving life and the second law of thermodynamics, and accounts for evolving systems of complexity as the very product of the intrinsic nature of the universe in all steps from the formation of atoms to the peak forms of life and humanity, and presumably beyond. Aharon Katchalsky has presented some interesting details on living systems as dynamic dissipative flow systems which, by the intrinsic nature of things, evolve in similar stepwise fashion. Herbert A. Simon's "The Architecture of Complexity"[14] provides other details on understanding cultural and even computer evolution in a similar context. B. F. Skinner in his "Phylogeny and Ontogeny of Behavior"[15] provides an illumination of ontogeny or individual development in a similar frame.

Values within this picture of evolving systems of life are made up of the complex hierarchy of physical forces that establish and maintain the norms that maintain the living system in being and allow it to take part in further evolution. One must note that the system of human values involves a complex hierarchy of subvalues that are interrelated all the way from atoms to complex molecules, to cells, to organs, to organisms, to societies, to ecosystems. The values necessary for this complex hierarchy are encoded in each of these levels in different ways, and the expression of these codes of information or memory is shaped by all of them acting collectively. There are cybernetic or goal-directing mechanisms in families and societies and species that integrate with the total requirements of the human ecological niche.[16] A man's behavior, conscious and unconscious, social and private, is motivated by a hierarchical network of forces that involve the cumulative memory patterns of his life history, of his culture insofar as he has experienced it, of his genotypic sample from the human gene pool, and ultimately of all these as they have been created and adapted in a series of habitats to form an ecological niche in the world or cosmic system. While some writers limit their use of the term "values" to some fragments of this total hierarchical network of forces—such as to conscious intent, or to social values, or to future values—I think a system dynamics view suggests we would introduce error and confusion if we cut off any part of this system of forces that shapes the human behaving that produces human life.

I have presented a very "materialistic" picture of human values. Yet I assert that this model applies as much to the highest flights of human aesthetic, spiritual, and religious experience, as it does to maintaining body temperature at about 98° F. The machinery is all of a piece, so far as the central nervous system is concerned. The "materialistic" picture of human values will not disturb those who have understood my earlier note on the equivalence of subjectivity and objectivity.

But we should encompass in our picture of human values the sources outside of the head from which the values come. The genotype is a model or system of the world that has been selected by that world in a course of development for over a billion years. The machinery of the general computer in the central nervous system, and the hierarchy of norms or basic values preset or acquired in that system, are provided by information that the larger world system has selected and embodied in the DNA of the genotype. All that any individual does must in the first place conform to that model or system of the world in his genotype and to the boundary conditions or information set by the local environment. Nothing lives apart from meeting such requirements.

The cultural heritage or culturetype must, like the genotype, provide a statistically valid set of norms that can match or integrate into a statistically significant portion of the genotypes in the gene pool with which it is involved, as well as into a statistically significant portion of the events likely to occur in the habitat of the population. The storehouse of information accumulated in a culture is not so much of a mystery today as it used to be. The memory banks of the central nervous system of the population, possibly in RNA instead of DNA form, may be the primary recording site of cultural information.[17] Since the emergence of linguistic symbol systems, insemination or transmission of the culturetype is in large measure by mouth and ear. What is of significance about this picture of the storage and transmission of values is that human behavior and human values are a part of the mechanisms of what has been called the material world, and that the events of the human spirit are as tied to determination by fate or by the invisible spirits (entities composed of protons and electrons) as they were ever thought to be in ancient religions. Human values and human behavior (including those of which we are aware) are no more exempt from expression in genetic, neural, or other records of information than the values and behaviors of anything else. Human values and behavior may, like other phenomena, be processed in logic, equations, and computers, as well as inside the head. There is not more than a ghost of a chance of human values being a separate or independent "world system," as many believe.

The cultural storehouses of information provide input into the brain of each developing child to shape his norms or goals. A general systems approach would seek to show how the structures or behavioral patterns of a society as observed by anthropologists and psychologists and other students of behavior are related to the patterns inside the central nervous systems, and how the products of interactions among the varied individuals constituting a society (each with a uniquely different genotype and each with a uniquely different developmental history in a subhabitat of the general environment) operate to transmit such homeostatic or steady state systems as languages, religions, technologies, economies, etc., and how all these relatively stable patterns of information integrate with the information supplied by the genotypes and the environing boundary conditions of the nonhuman world system so as to provide dynamic homeostasis for the world system and the total ecological niche of Homo.

The various fields of science have produced some theoretical or conceptual models for understanding these complex processes. I think we have enough established information to warrant our talking about a selection process operating on all levels of our hierarchic organization, and I have written about this in more detail elsewhere.[18] Thus I would say that human social values—even when they are generated within a person by cultural inheritance, or by immediately apprehended or intuited inspiration, or by elements from the genetic core of information that structures Homo sapiens differently from all other species on earth—have been arrived at by a process of natural selection—selection by the nature of what constitutes stable states inherent in the system, i.e., in the world system being revealed by the sciences today.

Of course, for people not acquainted with a broad range of the spectrum of the scientific pictures of man, human values may be described in more conventional language, and it is perfectly all right and in fact rather necessary to talk to each person in his own language. But for seeking to understand human values in the context of the new world system, and seeking to adapt our values to it, we need to use the language or model appropriate to that system. That language or symbol system is furnished by the sciences.

THE WORLD SYSTEM OF JAY FORRESTER AND HUMAN VALUES

I turn now to the world system of Jay Forrester and its relation to human values. Forrester, I believe, is using the same general or scientific conceptual system of the world, and of men within it, that I am using. His model is limited to providing sufficient information to predict outcomes of some of the rapid changes taking place, as Homo sapiens exploits a new ecological niche provided by the cultural mutation we call "modern scientific technology."[19]

Since Forrester's predictions of the future of the world system involve computers, I should insert a note about the involvement of computers in our models of world systems. It is quite conceivable to me that some kind of computer mechanisms may evolve faster to higher levels of understanding and power for modeling life than men possibly can, and I would refer you to my paper on "Evolving Cybernetic Mechanisms and Human Values"[20] for more details. It may be objected that computers cannot do more than operate according to their programs, but as some computer experts have pointed out, this is also true of man, who is programmed by his genotype, culturetype, and the general environment.[21] The point is that self-maintaining and self-reproducing computers may not only become more and more essential as "symbiotic" supports for man's entry into an increasingly scientific and technological society whose full dimensions we have only begun to see in this century, but computer evolution may become, under certain foreseeable conditions, independent of and more viable than Homo. We should be no wiser to brush aside this potential of computers than an interplanetary visitor a billion years ago would have been, had he said there was no significant future for the crudely organized blobs of complex chemical molecules making up the coazervates that were ancestors to living cells.

I believe that Forrester is quite right in his analysis that some of our intuitive human projections are likely to be wrong, partly because human brains do not have the same capacity as computers for manipulating a large number of variable functions in their models of the world. I would agree with critics of Forrester that he or some of his defenders sometimes overlook the degree to which computer models fail to correspond to the complexities of the "real world" and thereby project erroneous conclusions. But I don't think there is any real quarrel here. The people who are making models of world systems for computer operations to project the future are not really in a different situation from any of us who want to speak of the future. We all have to test our capacity to handle the variables, the validity of our symbolic calculus or logic, and the validity of our models of the initial states in the process of producing the subsequent states of our experience. I think that we must accept at face value the claims of the Club of Rome and the MIT System Dynamics Group that they are only seeking better information models of future states in order to avert catastrophe. We can be grateful to them for their efforts. Their world system is an important contribution to ours.

I want to give sharper focus to one element of their world system about which many apparently have not yet heard: man's long-term values. Dennis Meadows and Forrester have both written on this matter. Meadows has pointed out that *"the real constraint on global simulation* is that we lack the unifying theories of global evolution which indicate to us how it is that technology, population, and [human] *values* interact."[22]

I want especially to call attention to Jay Forrester's article on "Churches at the Transition between Growth and World Equilibrium," presented to the Division of Overseas Ministries of the National Council of Churches on November 4, 1971.[23] Here Forrester paints a system dynamics picture of the nature and role of religion.[24] Forrester's picture focuses on religion's role in maintaining the *long-term* values necessary for the survival of a society, and religion's special role now that human cultural evolution is shifting gears from its expanding or radiative exploitation of a new ecological niche (brought about in recent centuries by the emergence of scientific tech-

178

nology) into a new level of stable state or homeostasis within the boundary conditions given by the new situation. Step or quantum-jump transitions are characteristic of all evolutionary emergents from one level to a new one. Exponential expansion is a characteristic of the times of the emergence of new organisms, new species, or new phyla. After the expansion or radiation, new patterns of dynamic equilibrium become effective until new breakthroughs occur.

Forrester takes very seriously religion and its role in shaping man's long-term values. He points out that "to the extent that religious teachings have influence and carry weight in social decision-making, those religious beliefs must be included in a model that explains the dynamics of a society. Religious beliefs interact with other decision-making influences in a social system and are a part of the total policy structure that may produce either good or evil. Ethical principles interact with the principles of economics, technology, sociology, agriculture, and medicine to create today's pressures and social stresses."[25] While there would be some scholars of religion who would object that a social-function interpretation is not the only way to interpret religion, I think there are few who would not recognize religion's function as an agent of the social order.

I say this in spite of the fact that today there is a vigorous doubting, by many social-science students of religion, that there are any significant connections between religious beliefs and human behavior.[26] The problem in the social-science study of religion, as I have analyzed it, is that much that goes on as a study of religious belief is really a study of fossil religious belief that today is only alleged, and which is different from living belief.

In any case, no one will doubt that there are agencies that provide man with his long-range values. The genotype itself has done this for hundreds of millions of years. That some of the cultural transmissions of information do this is equally certain; otherwise languages, technologies, and other cultural phenomena would not be relatively stable systems and would have disappeared rather than have remained in force for millennia. If we will agree that religion is an appropriate name for those elements of culture that do in fact transmit a certain class of man's long-range values, we can use this element of an operational definition to help make religion discussable scientifically.

Forrester in his paper on "Churches" says that "historically, religious values have probably developed in response to long-term social needs. . . . Without the long-term values, [then] 'living for the present,' if carried to an extreme, makes the future impossible. The societies that have grown and prevailed are those with a viable concept of the future. Without such an enduring values set, the society fails to develop, decays from the inside, or is replaced by a more future-oriented social system."[27] In my opinion, this is a useful and valid picture of religion and human society.

Forrester suggests: "The church should be custodian of the longest-term values in a society. Those values should look beyond civilian laws and national constitutions. As custodian of the future, the church should understand that long-term values will conflict with the short-term values and goals of man and society. The church must have effective ways to project long-term goals into the current processes of everyday decision-making."[28] At the same time, he notes the failure of the church to present credible and relevant values for today's society, and that "without an appropriate long-term value structure, the society begins to falter."[29] He points to the church's selling of its soul as the custodian of long-range values and espousing short-range secular values in order to save its own life in a disbelieving society. But, he says, "the church will not be effective if it has lost its distant time horizon. If its values are the same as those of the secular society, the church need not exist."[30] "The church, if it is to be the custodian of long-term values, must define right in terms of the enduring and future welfare of mankind. Other institutions will adequately defend the short run."[31]

I find Forrester's analysis of the church quite in keeping with views of various social-science studies of religion, and some theological apologetics. The "long-term values" are in some ways close to Paul Tillich's "ultimate concerns." Forrester even transcends many contemporary theologians in pointing to transhuman and transcendent virtues and powers when he writes: "Is selfishness any less sinful when exercised in favor of mankind as a whole than when for oneself?"[32] Here he is indicating that the short-term existing values within human nature may be fatal for the long-term existence of man as determined by superhuman realities on which man is totally dependent.

In summary, Forrester says that "from a systems-dynamics viewpoint, religious teachings are a part of the policy structure of a society."[33] Long-term values are a necessity, and a large part of our trouble today is the failure of religions to transmit suitable and convincing long-range values to men. As I suggested in my opening paragraphs, when we begin to analyze man's situation in terms of the ultimate ecosystem (our largest and longest world view), we are beginning to speak about the determiner of human destiny, about the ultimate sources and constraints that have created and sustain human nature. Since in a systems analysis no relevant part is excluded from the analysis of the whole, we find ourselves constrained to look into an aspect of man's nature—his ultimate concerns and the ultimate creator or determiner of his destiny—that has been explored in earlier stages of cultural evolution by the religious institutions. In response to Forrester's and Meadows's call for some better understanding of how man's long-range values are properly established, I want to turn now to some sugges-

RALPH WENDELL BURHOE

tions for a new synthesis of some elements of the scientific world system with some of the traditional religious models. My prediction is that widespread public behavior in accord with sound long-term values will await the revitalization of religion and that this is not likely to take place until the wisdom already evolved within the traditions is translated and interpreted within today's more extended and credible views of man and the world developed by the sciences. This holistic task of interpreting man's ultimate destiny in the scheme of things might well constitute the ultimate concern of systems philosophy. In any case, I think an analysis of man's ecological niche in the ecosystem may, for the first time since Copernicus and Galileo—when science and religion began to separate and split Western culture—make possible a reunion or new synthesis. I suggest the following.

A SYNTHESIS OF SOME TRADITIONAL CONCEPTS OF GOD WITH THE NEW CONCEPTS OF COSMIC ECOSYSTEM

1. I think we can concur with Forrester that religions, according to critical history and anthropology, as well as their own claims, may be said to be the institutions that are primarily concerned with man's overall and longest-range values. This is what "salvation of the soul on the last day" is all about. Since man is a social animal, the long-range values of religions have to be conceived as functioning to integrate the individual man into the social values or goals required for the viability of his society. These social values are not given man by his genotype in anything but their crudest tendencies or bases, and hence must be cultivated by social institutions.[34]

2. We can say that these long-range social and personal

values, like all other values, are as "real" and "objective" as anything else in a scientific world view. The *established* feedback patterns or cybernetic mechanisms, and the norms or goal settings of these mechanisms, can be described, and we can say something about their evolution and the reasons for their being what they are, just as we can explain the DNA patterns that produce the opsin biochemicals necessary for visual reception in the retina. The *still-to-be-discovered* norms, the potential values, also may be projected by the scientist and the computer as well as by the prophet and poet.

3. Traditional religious belief systems characteristically have involved man's relation or adaptation to some ultimate realities which vastly transcend man's power and whose requirements man must discover and obey if he is to live well or at all. Belief in such transcendent powers has provided motivation for good behavior, social and personal. While the popular interpretation of the new and expansive powers given to man by recent science has fostered a prideful feeling that the gods are dead and man is master of his own fate, with powers to do almost as he pleases, the basic scientific world view portrays the reverse. Man can fly only if he learns nature's requirements and obeys them. An analysis in terms of the total ecosystem in which man lives and moves and has his being shows (a) that the world ecosystem created man and man did not create himself, (b) that this world system sustains man and that man is utterly dependent upon such an ecosystem and can have no life at all if he violates the boundary conditions (inside as well as outside the system) which the system sets for reasonably steady states of the hierarchical network of open systems in which man participates. Within this scientific world picture, we move beyond freedom and dignity to a traditional theological view of man's utter dependence upon, and necessary service to, the will of the transcendent power. We also find that man is indeed made in the image of his creator, and that the

informational system in the heart of his soul, which properly motivates him to maintain and advance life, is a reflection of a very long-enduring and costly generation of an image of the ultimate cosmic system. The command, implicit in the cosmic program of emerging stable states as each new step on the ladder of evolution is reached, is a command to seek life and not death. We find man motivated and enabled to seek higher levels of life by a sacred history already inscribed in the core of his being. In this picture, man finds himself created by grace, nourished by grace, and saved by grace of a system far transcending himself and his knowledge.

4. Since a cardinal sin of man is to ascribe, in pride, to himself, instead of to the ultimate reality which is his creator and sustainer, the true or right definition of how to live, salvation can only come when man repents and recognizes that there is in fact a reality upon which he is totally dependent. To avoid psychological problems (or conflicting goals inscribed in his brain), cultural information must supplement genetic information about who man is. The obvious death of the body has, in the millennia since man became aware of death, pushed him to seek a model or image of a human being as more than his body. The religious solutions have been ingenious and are remarkably parallel to the scientific pictures. In scientific language, reincarnation of the genotype and the culturetype, or resurrection of the essential life pattern from the ashes after the death of the body—that is, genotypic and culturetypic continuity or immortality of the essential and enduring human core—are sharp new models of the models of reincarnation, resurrection, or immortality of the human soul in traditional religions. By better images of his own nature as a part of the larger world system which is his creator, and the evolving ecological niche which is a larger view of his soul, man can have proper long-range goals, including proper hope for ultimate triumph of good over evil and life over death. The

long-range goals then can become real, significant, and motivated—as they are *not* for a man who conceives that he should eat, drink, and be merry, for tomorrow he may die, or for the man who thinks he is not ruled by a transcendental reality. The hope—which an understanding of the long-range reality of the human soul in the hands of a creative and omnipotent God can bring to man—is a hope or goal which is necessary if we are to motivate the kinds of long-range human values that Forrester says are necessary. I suggest we can accomplish this by translating doctrines of an almighty God and a death-transcending soul of man into the symbols of the modern scientific world view.

5. With the resurrection of a God concept, and with the restoration of many other related concepts from the traditions of the higher religions—now possible in the light of the holistic views of a world systems analysis—there is a possibility that those semidormant and currently rather ineffectual institutions in human societies that have the function of transmitting man's ultimate concerns or long-term values may become revitalized and rise again in time to turn man from his selfish vanities and save him. The cultural wisdom embodied in these institutions[35] —a wisdom of which we do not yet have good scientific explanation—probably makes them the best instruments for providing man with the long-term values he requires for living in the new age. The main problem is to show that some of the prime elements of their wisdom are not—as a shallower analysis has implied—basically "unreal," but are validated by the hardest sciences, when their materialistic and deterministic doctrines are properly translated.

9

Comment

JAY W. FORRESTER

THE papers in this volume present fascinating, diverse, and informative viewpoints on how we may better model and thereby understand the world of which we are a part. With most of the comments I agree. Where I differ, misunderstandings seem to be the issue rather than incompatible philosophy.

Through the papers runs a persistent uneasiness about the apparent absence of psychological, sociological, and political variables in models such as that in my *World Dynamics* and the one described by Meadows in *Limits to Growth*. I say there is only an apparent absence because, at the implicit level, such variables are strongly present. In the models, the humanists miss a connecting linkage in terms of human values and psychological forces just as the economists miss a price system. But both changing values and changing prices are intervening variables in real life that connect the nature of the world to human reactions. They are not missing from the model but are swept up in the high degree of aggregation until their detailed terminology is submerged.

We see in the United States in this decade a falling birth rate. Some have suggested that the falling birth rate lies outside the scope of the present models and is to be attributed to the social and psychological variables that have been omitted. But from whence come the influences leading to the social and value changes? Are they not from the sense of crowding, from the material standard of living that is now absurdly high for a substantial percentage of the population, from the highway congestion, from the pollution, and, in short, from the pressures that are being reflected back from the natural barriers as population and industrialization begin to impinge on the limits of our surroundings? I suggest that psychological attitudes and social norms are shaped by physical circumstances and are the intervening variables between the condition of the world and the human responses. In the *World Dynamics* model, birth rate falls as crowding increases; the numerical values are such that the effect clearly does not come from physical crushing; the effect is primarily psychological. The effect of food on population need not be physical starvation, it can be the threat of hunger.

The issue here is more methodological than philosophical. One must always compromise between simplicity and completeness in constructing models. There is no right answer except in the context of the purpose of the particular model. The world models have focused on the broad sweep of major forces. At a later stage it will be appropriate to insert more connecting tissue. This is not to suggest that details of the connecting tissue will have no effect. The model in *World Dynamics*, by omitting the level variables that represent human values and attitudes, is not omitting such variables but is saying that the delays in accomplishing value change can be neglected for the particular purposes. Were those delays explicitly inserted, additional dynamic interactions would emerge. The changes would probably lead to worsening an already forbidding glimpse of the future. Far from providing a solution to the problems of mankind, the insertion of the psychological and political delays between the world condition and the human response would lead to more overshoot of population beyond the carrying capacity of the globe.

Psychological and social variables have not been omitted. They are subsumed in the variables already present. To the extent that we care to state a hypothesis about such intervening variables, they can be readily included in systems dynamics models. Such has been done in models more complex than represented in *World Dynamics*.

190

Another concern expressed about the world models relates to the aggregation of real variables. For example, the developed and underdeveloped nations are not separated. Here again the answer lies in purpose. If one were dealing with the relative struggle between the two groups, or if he were examining whether or not the income gap would be closed, the two would need to be separated and the differential forces between them would need to be represented. However, if the emphasis is on the total loading of the environment by the total population of the world and its total capital plant, then the issues arise not from the distinctions between but from the sum of the two. One must understand the models well enough to know the areas for which they are usable. Every practical model will have limits beyond which it is not useful. The proper attitude toward the world models is to look for what they can teach us and identify areas of inapplicability only so that they will not be used for the wrong purposes.

I was especially interested in the changing political perspectives outlined by Alastair Taylor as they have evolved in response to changing technology of power. The scope of power has gradually expanded from the family to the tribe, city, country, and now to the national alliances. But what does an understanding of the world system lead to? Will it be "one world" and an effective world government, or will the pendulum swing back to the independent nation?

At what level is the compromise to be made between population density and nature? We face the trade-off between quantity and quality. Every country is capable of supporting a sufficiently small population at a high standard of living and quality of life. But one of the freedoms is the freedom to choose between the size of the population and the conditions under which that population lives. Is the compromise to be made by a world authority that imposes the same balance on every culture? If not the choice must be decentralized. If the

191

choice is made differently in various countries, then the standard of living cannot be the same. Some countries will choose a modest population coupled with a high material standard of living, national strength, and the disadvantage of maintaining the self-discipline to limit population. Other countries will, in effect, choose to avoid the trauma of self-discipline or perhaps belong to cultures that prefer a higher population density and will accept the corresponding reduction in material standard of living and in national strength.

Such reasoning suggests that each country must live within its own capability to a greater extent than today. The present accelerating pace of international trade is a device to allow growth to continue until the entire world simultaneously approaches shortages of all traded goods. Then we are apt to see hoarding for the future by resource-supplying nations and a consequent contraction of economic activity and standard of living in those countries that have expanded beyond their internal means.

The urgent task now is to face such issues squarely and to make an estimate of the most viable and realistic future. Much of today's actions are based on visions of impossible future utopias. Unrealistic expectations are a poor foundation on which to build the future. Hard choices must be made. The future is not to be free of pressures, but we have a range of choice in the combination of physical, social, psychological, and moral pressures under which we will live. Systems dynamics modeling can cope with such considerations as rapidly as we can think through the issues and identify the important relationships. From such an effort it will be possible to answer many of the questions raised in this book.

Notes

1. *Uses and Misuses of World System Models*

1. Jay W. Forrester, *World Dynamics* (Cambridge, Mass.: Wright-Allen Press, 1971).

2. D. H. Meadows, D. L. Meadows, W. W. Behrens III, J. Randers, *The Limits to Growth: A Report for the Club of Rome's Project on the Predicament of Mankind* (New York: Universe Books, 1972).

3. Jay W. Forrester, "Behavior of Social Systems" in *Hierarchically Organized Systems in Theory and Practice*, Paul A. Weiss, ed. (New York: Hafner, 1971).

4. Jay W. Forrester, "Churches at the Transition Between Growth and World Equilibrium," *Zygon*, vol. 7, 3 (September 1972). See Burhoe, pp. 178 f., below.

5. Meadows, *et al.*, op. cit.

6. Edward Goldsmith, Robert Allen, Michael Allaby, John Davoll, and Sam Lawrence, "A Blueprint for Survival," *The Ecologist*, vol. 2, no. 1 (January 1972).

7. Paul A. Weiss, "The Basic Concept of Hierarchic Systems," *Hierarchically Organized Systems in Theory and Practice*, op. cit.

8. Cf. Ervin Laszlo, "World System Research and Information Bureau: A Proposal," paper read before the annual meeting of the Society for General Systems Research and the American Association for the Advancement of Science, December 29, 1972, Washington.

2. *Models and Systems Analyses as Metacommunication*

1. G. Gordon Brown, Margaret Mead, and Eliot D. Chapple, "Report of the Committee on Ethics," *Human Organization*. vol. 8 no. 2. (Spring 1949), pp. 20–21.

2. D. H. Meadows, et al., *The Limits to Growth* (New York: Universe Books, 1972).

3. Gregory Bateson, *Steps To An Ecology of Mind* (San Francisco: Chandler Publishing Co., 1972).

Mary Catherine Bateson, *Our Own Metaphor* (New York: Alfred A. Knopf, 1972).

3. *Some Political Implications of the Forrester World System Model*

1. Alastair M. Taylor, "The Computer and the Liberal: Our Ecological Dilemma," *Queen's Quarterly,* vol. LXXIX, no. 3 (autumn 1972), p. 293.

2. Though not to the passing of the West's once unchallenged global political and economic hegemony, as witnessed by post–World War II decolonization affecting over a billion people. In *The Great Frontier,* Walter Prescott Webb has sought to incorporate Turner's American Frontier thesis in this larger historico-geographical framework of European expansion from the fifteenth century to its conclusion during our own.

3. It should be pointed out that one-to-one relational thinking is a special case of two-termed or dyadic relationality. The character of a one-to-one relation applies to the *fields* of two relations, as shown by Bertrand Russell in his *Introduction to Mathematical Philosophy,* chapter 6.

4. To list the criteria set forth in the Montevideo Convention of 1933 on the Rights and Duties of States, Article I.

5. Oswald Spengler, *The Decline of the West,* abridged ed., trans. Charles F. Atkinson (London: George Allen and Unwin, 1961), pp. 89, 90.

6. Karl Ritter, *Comparative Geography* (Philadelphia: J. B. Lippincott, 1864), p. 184.

7. Friedrich Ratzel, *Politische Geographie* (1897).

8. I have discussed this question at greater length in *Integrative Principles of Modern Thought,* Henry Margenau, ed. (New York: Gordon and Breach, 1972).

9. Hans J. Morgenthau, *Politics Among Nations,* 5th ed. (New York: Alfred A. Knopf, 1972), chap. 1.

10. "Whenever peace—conceived as the avoidance of war—has become the primary objective of a power or group of powers, international relations have been at the mercy of the state willing to forego peace. . . . Peace, therefore, cannot be aimed at directly; it is the expression of certain conditions and power relationships." *Nuclear Weapons and Foreign Policy* (New York: W. W. Norton, 1969), p. 244.

11. "The Geographical Pivot of History," *The Geographical Journal,* vol. XXIII, no. 4 (1904).

12. Mackinder's "Pivot Area" as described in 1904 fell largely within the Russian state. In his *Democratic Ideals and Reality* (1919), he renamed it the "Heartland" (with its two crescents); viewing Eastern Europe as the key, he enunciated his well-known dictum:

> Who rules in Eastern Europe commands the Heartland,
> Who rules the Heartland commands the World-Island,
> Who rules the World-Island commands the World.

This dictum inspired the interwar German geopoliticians. An American political geographer, Nicholas Spykman, however, contended that Mackinder had overemphasized the Heartland while insufficiently appreciating the crucial strategic role of the inner crescent—which he called the "Rimland"

—that formed a vast buffer zone of conflict between sea and land power. Who controlled this Rimland ruled Eurasia and therefore controlled the destinies of the world. *The Geography of Peace* (New York: Harcourt, Brace, 1944).

13. A. P. de Seversky, *Air Power: Key to Survival* (New York: Simon & Schuster, 1950).

14. See Alfred North Whitehead, *Science and the Modern World,* chap. 6, "The Nineteenth Century"; John Herman Randall, Jr., *The Making of the Modern Mind;* and Floyd W. Matson, *The Broken Image,* chap. 1, "The Mechanization of Man."

15. Emer de Vattel, *The Law of Nations, or the Principles of Natural Law Applied to the Conduct and to the Affairs of Nations and of Sovereigns* (1758).

16. Quincy Wright, *A Study of War,* abridged ed. (Chicago: University of Chicago Press, 1964), p. 135.

17. *Ibid.,* p. 134.

18. Richard Hartshorne, "Morphology of the State Area: Significance for the State," in C. A. Fisher, ed., *Essays in Political Geography* (London: Methuen, 1968), p. 27. His theory is set forth in "The Functional Approach in Political Geography," found in the *Annals* of the Association of American Geographers, vol. XL (1950).

19. "The Functional Approach in Political Geography," op. cit.

20. Friedrich Ratzel, *Politische Geographie,* 3d ed. (Munich and Berlin: R. Oldenbourg, 1923), p. 6.

21. Jean Gottmann, "The Political Partitioning of Our World: An Attempt at Analysis," *World Politics,* vol. IV, no. 4 (1952), pp. 512–519. As a counter centrifugal force, Gottmann advanced the concept of circulation, based on the argument that technological developments lead to improved circulation or mobility, and hence to change.

22. That the contemporary interaction of technological and other forces makes for what he calls "supranational integration and equalization of nations" is the thesis of Silviu Bracan in *The Dissolution of Power* (New York: Knopf, 1971).

23. See Edward W. Soja, *The Political Organization of Space,* Association of American Geographers, Washington, D.C. (1971), pp. 28 ff. For my part, I would apply the principle of integrative levels in order to recognize the inputs of our "biological level" of organization in our physiological perception of space and our territorial apperceptions in turn, while also agreeing with Dr. Margaret Mead that our cognitive-affective processes—as conditioned by socialization and culture change—are the fundamental determinants in man's relations with his environment and fellow man.

24. Soja, *op. cit.*

25. Alastair M. Taylor, "Toward a Field Theory of International Relations," *General Semantics Bulletin* 35 (1968), p. 19. On changing conceptions of physical space, see Max Jammer, *Concepts of Space,* 2d ed., (Cambridge: Harvard University Press, 1969); Hans Reichenbach, *The Philosophy of Space and Time* (New York, Dover, 1958). As regards cur-

rent geographical literature, consult R. M. Downs, "Geographic Space Perception: Past Approaches & Futurative Prospects," in Board et al., eds., *Progress in Geography,* vol. 2 (London: Arnold, 1970); B. Gooday, *Perception of the Environment: An Introduction to the Literature,* University of Birmingham Center for Urban and Regional Studies, Occasional Paper no. 17 (1971); and T. F. Saarinen, *Perception of Environment,* Association of American Geographers, Resource Paper no. 5, Washington D.C. (1969).

26. Harold and Margaret Sprout, "Environmental Factors in the Study of International Politics," *The Journal of Conflict Resolution,* I, no. 4 (1957), pp. 309–328. See also *An Ecological Paradigm for the Study of International Relations* (Princeton University Press, 1968). For a study of the relation of spatial perceptions to cartographical projections, see Waldo R. Tobler, "Geographic Areas and Map Projections," B. J. L. Berry and D. F. Marble, eds., *Spatial Analysis* (Prentice Hall, 1968), pp. 78–90.

27. Stephen B. Jones, "A Unified Field Theory of Political Geography," *Annals* of the Association of American Geographers, vol. XLIV (1954).

28. See Karl W. Deutsch, *Political Community at the International Level: Problems of Definition and Measurement* (Princeton University Press, 1953); also "The Growth of Nations: Some Recurrent Patterns of Political and Social Integration," *World Politics,* V (January 1953), pp. 168–195.

29. Jones, *op. cit.*

30. R. E. Kasperson and J. V. Minghi, eds., *The Structure of Political Geography* (Chicago: Aldine, 1969), p. 12.

31. A special volume of *International Organization,* vol. XXV, no. 3 (summer 1971) entitled *Transnational Relations and World Politics* contains articles in these specific areas and also excellent introductory and concluding chapters by its editors, Joseph S. Nye, Jr., and Robert O. Keohane.

32. See, for example, Wendell Bell and Walter Freeman, eds., *Ethnicity and Nation-Building: Local and International Perspectives* (Sage Publications, 1972); and L. G. E. Edmondson, "Africa and the African Diaspora: Interactions, Linkages and Racial Challenges in the Future World Order," forthcoming in A. A. Mazrui and H. H. Patel, eds., *Africa and World Affairs: The Next Thirty Years* (New York: Third Press).

33. As a pioneering venture in this area, see G. A. Almond and J. S. Coleman, eds., *The Politics of the Developing Areas* (Princeton University Press, 1960).

34. Ernst B. Haas, "The Study of Regional Integration: Reflections on the Joy and Anguish of Pretheorizing," *International Organization,* vol. XXIV, no. 4 (autumn 1970); this issue is given over to *Regional Integration: Theory and Research* and carries articles by integration specialists, including Lindberg, Puchala, Nye, Alker, Hayward, and Scheingold.

35. B. B. Hughes and J. E. Schwarz, "Dimensions of Political Integration and the Expérience of the European Community," *International Studies Quarterly,* vol. 16, no. 3 (September 1972), pp. 263–294.

36. See, for example, G. A. Almond's "A Functional Approach to Comparative Politics" in the aforementioned *The Politics of the Developing*

Areas; Morton Kaplan, "The Systems Approach to International Politics" in his edited work, *New Approaches to International Relations* (New York: St. Martins, 1968); and J. Rosenau, *The Adaptation of National Societies: A Theory of Political System Behavior and Transformation* (New York: McCaleb-Seiler, 1970).

37. We might cite C. A. McClelland's two works, "Applications of General Systems Theory," in J. Rosenau, ed. *International Politics and Foreign Policy* (New York: Free Press, 1961), and *Theory and the International System* (New York: Macmillan, 1966); P. Nettl, "The Concept of System in Political Science," *Political Studies,* vol. XIV, no. 3 (October 1966); D. Easton, *A Framework for Political Analysis* (Prentice-Hall, 1965); and J. W. Burton, *Systems, States, Diplomacy and Rules* (Cambridge University Press, 1968). For an application of positive feedback to political systems, see C. R. Dechert, "Integration and Change in Political and International Systems," *Positive Feedback: A General Systems Approach to Positive/Negative Feedback and Mutual Causality,* J. H. Milsum, ed. (Pergamon, 1968).

38. For one such incisive criticism, see J. Stephens, "An Appraisal of Some System Approaches in the Study of International Systems," *International Studies Quarterly,* vol. 16, no. 3 (September 1972), pp. 321–349.

39. David Harvey, *Explanation in Geography* (London: Arnold, 1969), p. 468.

40. See R. J. Chorley and B. A. Kennedy, *Physical Geography, A Systems Approach* (London: Prentice-Hall, 1971); also R. J. Chorley, "Models in Geomorphology," *Models in Geography,* R. J. Chorley and P. Haggett, eds. (London: Methuen, 1967).

41. R. Abler, J. Adams, and P. Gould, *Spatial Organization* (Englewood Cliffs: Prentice-Hall, 1971).

42. J. B. McLoughlin and J. N. Webster, "Cybernetic and General System Approaches to Urban and Regional Research: A Review of the Literature," *Environment and Planning,* 2 (1970), pp. 369–408.

43. E. W. Soja, "A Paradigm for the Geographical Analysis of Political Systems," paper presented at the annual meeting of the American Political Science Association, New York, 1969.

44. S. B. Cohen and L. D. Rosenthal, "A Geographical Model for Political Systems Analysis," *Geographical Review,* vol. 61, no. 1 (January 1971), pp. 5–31. In their readings in political geography, Jackson and Douglas have adopted a systems perspective, based loosely on the Eastonian framework. However, the selection of articles comprises for the most part ideographic case studies which in themselves do not develop a comprehensive, far less a rigorous, systems approach to the discipline: cf. W. A. D. Jackson and M. S. Samuels, eds., *Politics and Geographic Relationships,* 2nd ed. (Englewood Cliffs: Prentice-Hall, 1971).

45. G. A. Almond and G. B. Powell, Jr., *Comparative Politics: A Developmental Approach* (Boston: Little Brown, 1966), p. 25.

46. *A Framework for Political Analysis* examines the theoretical status of a political system; its environment; systemic persistence in a world of stability and change; the political system under stress; and types of regulative

responses. See also *A Systems Analysis of Political Life* (New York: John Wiley & Sons, 1965).

47. "In this sense outputs are exemplified in the statutes of a legal system, administrative decisions and actions, decrees, rules, and other enunciated policies on the part of the political authorities. . . . Just as inputs are a way of organizing and communicating the effects of environmental changes to the political system, outputs reverse the process. They represent a method for linking up what happens within a system to the environment through the unique behavior related to the authoritative allocation of values." *A Framework for Political Analysis,* pp. 126–127.

48. *Ibid.,* p. 128.

49. Knowledge (information) is always subject to loss, especially when transmitted societally rather than genetically; for more on the application of the principle of integrative levels to sociocultural systems, see the author's chapter in *Integrative Principles of Modern Thought,* Henry Margenau, ed. (New York: Gordon and Breach, 1972).

50. The diagram above is meant only to map the application of the principle of integrative levels in its most elemental structure: each of the levels could be further differentiated into sublevels—either as specific societal and/or political systems, for example, or into stages of environmental control (such as S_4, which lends itself to further analysis of thalassic, oceanic, and continental stages of societal organization).

51. I borrowed these terms from Ervin Laszlo's "theory of natural systems," in which he describes "System-cybernetics I" as "function of adaptation to environmental disturbances resulting in the *reestablishment* of a previous steady state in the system"; and "System-cybernetics II" as "function of adaptation to environmental disturbances resulting in the *reorganization* of the system's state, involving, with a high degree of probability, an overall *gain* in the system's *negentropy* and *information content*." See his *Introduction to Systems Philosophy* (New York: Gordon and Breach, 1972; Harper Torchbooks 1973), p. 36.

52. Alastair M. Taylor, "Evolution-Revolution, General Systems Theory, and Society," in *Evolution-Revolution: Patterns of Development in Nature, Society, Man and Knowledge,* R. Gotesky and E. Laszlo, eds., (New York: Gordon and Breach, 1971), p. 123. As I stated therein, while overall sociocultural development has been "upward," i.e., in the direction of progressive complexification and heterogeneity, history is replete with instances of quantization from more complex to simpler levels of societal organization.

4. *Reforming World Order: Zones of Consciousness and Domains of Action*

1. April 28, 1971 (Release of United Nations Information Office).

2. Of the word, Herman Kahn has, with I suppose unwitting irony, aptly written, "This is an ugly word, smacking of pseudoscience, but we seem to be stuck with it." Kahn and B. Bruce-Biggs, *Things to Come: Thinking about the 70's and 80's* (New York: Macmillan, 1972), p. 1.

3. Cf., e.g., Thomas Stritch, "The Banality of Utopia," *Review of Politics*, vol. 34 (January 1972), pp. 103–6; Margaret Mead, "Towards more Vivid Utopias," *Science*, vol. 126 (November 1957), 957–61.

4. Herman Kahn and Anthony J. Wiener, *The Year 2000: A Framework for Speculation on the Next Thirty-three Years* (New York: Macmillan, 1967); for critical assessments see Marion J. Levy, Jr., "Our Ever and Future Jungle," *World Politics*, vol. XXII (January 1970), pp. 301–27; William Irwin Thompson, *At the Edge of History* (New York: Harper and Row, 1971), pp. 113–23.

5. The most celebrated, but by no means the only, product of this project has been its Report for the Club of Rome: Donella H. Meadows and others, *The Limits to Growth*, (New York, Universe Books, 1972); underlying this work has been the approach to the analysis of social systems developed in the work of Jay Forrester, see esp. Jay W. Forrester, *World Dynamics* (Cambridge, Mass.: Wright-Allen Press, 1961). The Project Director, Dennis L. Meadows, has provided important intellectual leadership as the prime interpreter of Forrester's work and as the leader of an effort to evolve an adequate computer model for the study of world dynamics. In my view the Forrester-Meadows undertaking, unlike that of Herman Kahn, is of genuine and enduring importance. It undertakes to evolve and test by the best data available a program for the most accurate possible computer model of world dynamics, and in that sense, is a cumulative enterprise whose eventual contribution to human understanding cannot yet be assessed. My criticisms are directed mainly at the effort to gain authority in public thinking for its analysis of the world situation on the basis of numbers and the use of a computer modeling technique. Their computer printouts provide nothing that is more authoritative than a qualitative argument at present of the sort available from many other authors who adopt an ecological view of the planet's drift into a position of irreversible jeopardy. See, e.g., Paul R. Ehrlich and Ann H. Ehrlich, *Population, Resources, and Environment*, 2d ed. (San Francisco: W. H. Freeman & Co., 1972); Barry Commoner, *The Closing Circle: Nature, Man, and Technology* (New York: Knopf, 1971); Harold and Margaret Sprout, *Toward a Politics of the Planet Earth* (New York: Van Nostrand-Reinhold, 1971); G. Tyler Miller, Jr., *Replenish the Earth: A Primer on Human Ecology* (Belmont, California: Wadsworth Publishing Co., 1972).

6. Among the backlash literature see Carl Kaysen, "The Computer that Printed Out W*O*L*F*," *Foreign Affairs*, vol. 50 (July 1972), pp. 660–68; Rudolf Klein, "Growth and Its Enemies," *Commentary* (June 1972), pp. 37–44; and see front page of the *New York Times* Sunday review of the Meadows-Forrester books by Peter Passell, Marc Roberts, and Leonard Ross, *New York Times Book Review*, April 2, 1972, pp. 1, 10, 12–13.

7. These criticisms are particularly well formulated in the essay by Kaysen, cited in note 6.

8. However, there is something to be said for sounding an ecological alarm, by whatever means, at this time. It is partly a matter of shouting "Fire!" in a theater attended by deaf mutes. And in this sense the Club of

Rome, Jay Forrester, Dennis Meadows must be regarded as lifeguards of prime importance. It should also be noted that Jay Forrester and Dennis Meadows are sensitive to possible charges of pseudoscience and are careful in their statements about the present limits of their model.

9. See book cited note 4; cf. also Thompson, "Planetary Vistas," *Harpers* (December 1971), pp. 71–78; Thompson, "The Individual as Institution: The Example of Paolo Soleri," *Harpers* (September 1972), pp. 48–62.

10. The most coherent formulation of Peccei's outlook is contained in his book *The Chasm Ahead* (New York: Macmillan, 1969).

11. Interview with Thompson, *Time*, August 21, 1972, p. 51.

12. In this spirit Thompson has praised the intellectual dialogue between the Indian mystical philosopher, Gopi Krishna, and the German physicist-philosopher, Carl Friedrich von Weizsächer. This dialogue is described by von Weizsächer's long introduction to Gopi Krishna's book *The Biological Basis of Religion and Genius* (New York: Harper and Row, 1972). Cf. also Thompson's assessment of the architect, Paolo Soleri, cited note 9, and Thompson's own conception of a learning/living center based on the ideas of the medieval Irish monastery at Lindisfarne and outlined in his prospectus for a Lindisfarne Association.

13. That is, the dichotomy between future goals and present responses suggests a form of distancing that is disturbing in its human consequence in the same way as the more concrete circumstances of nonresponse disclosed in relation to Kitty Genovese's murder before the eyes of her passive neighbors. At issue, also, is the matter of personal responsibility, its extent and character. In this sense, I believe that responsibility to act to stop crimes of war of the sort embodied in the Indochina War extends to all human beings. It is the affirmation of this universal bond of involvement and accountability that is the most significant outcome of the Nuremberg judgment, and of what has more recently come to be called the Nuremberg obligation by American citizens who have been acting in opposition to the continuing American involvement in the Indochina War in an organization called *Redress*.

14. See, e.g., V. R. Potter, *Bioethics: Bridge to the Future* (Englewood Cliffs, N. J.: Prentice-Hall, 1971).

15. The Seneca quotation is borrowed from Bradbury K. Thurlow in a monthly newsletter distributed by a Wall Street brokerage firm, Laidlaw & Co., *Commentary*, November 1972, p. 1 (investment newsletter).

16. "In a Manner that Must Shame God Himself," *Harpers*, Nov. 1972, pp. 60–68, p. 60.

17. Same page; "The Winners are rehearsing for *Things to Come*," p. 68. In a different vein, a similar kind of assessment of the future is presented by Michael T. Klare in his book, *War Without End: American Planning for the Next Vietnams* (New York: Knopf, 1972).

18. E.g., Chester L. Cooper, *The Lost Crusade: America in Vietnam* (New York: Dodd, Mead, 1970); Roger Hilsman, *To Move a Nation* (Garden City, N. Y.: Doubleday, 1967); Townsend Hoopes, *The Limits of Inter-*

vention: An Inside Account of How the Johnson Policy of Escalation was Reversed (New York: David McKay, 1969); Arthur M. Schlesinger, Jr., *The Bitter Heritage: Vietnam and American Democracy, 1941–1966* (Boston: Houghton Mifflin, 1966).

19. See Stephen Hymer, "The Multinational Corporation and the Law of Uneven Development," in Jagdish N. Bhagwati, ed., *Economics and World Order: From the 1970's to the 1990's* (New York: Macmillan, 1972), pp. 113–140.

20. See F. H. Hinsley, *Power and the Pursuit of Peace* (Cambridge, England: Cambridge University Press, 1963); Walter Schiffer, *The Legal Community of Mankind* (New York: Columbia University Press, 1954).

21. Within national societies laissez-faire has been a *policy* where it has prevailed, but within international society it is an inherent *condition*.

22. Grenville Clark and Louis B. Sohn, *World Peace Through World Law*, 3d ed. (Cambridge, Mass.: Harvard University Press, 1966).

23. On this point compare Kenneth Boulding, "The Prevention of World War III," in Falk and Saul H. Mendlovitz, eds., *The Strategy of World Order: Toward a Theory of War Prevention* (New York: World Law Fund, vol. 1, pp. 3–13, 1966), with Robert Osgood and Robert W. Tucker, *Force, Order, and Justice* (Baltimore, Md.: Johns Hopkins Press, 1967).

24. See, e.g., in addition to works cited in note 5, the manifesto of the editors of the British magazine *The Ecologist* published under the title *Blueprint for Survival* vol. 2, no. 1 (Jan. 1972).

25. This reliance is exhibited in international policy by placing stress upon alliances, by intervening in foreign societies to assist the efforts of sympathetic elites to retain or acquire power in struggles against potentially hostile elites, and by ideological rationalizations that convey to one's own population a higher motive than the maintenance of position in the structure of international power, wealth, and prestige. These ideological mystifications often function in such a potent way as to entrap the power wielders in their claims, thereby inducing some poor calculations based on "principles" rather than "interests." Henry Kissinger's critique of pre-Nixon foreign-policy-making during the Kennedy/Johnson presidencies rests on this kind of assertion. See, e.g., Henry A. Kissinger, "Central Issues of American Foreign Policy," in Kermit Gordon, ed., *Agenda for the Nation* (Washington, D.C.: The Brookings Institution, 1968), pp. 585–614.

26. Jay W. Forrester, "Churches at the Transition Between Growth and World Equilibrium," *Zygon*, 7 (3) p. 146.

27. The approach in the remainder of this chapter is much more fully developed in chapters III and IV of the final document of the North American Section of the World Order Models Project (available in mimeographed form from the World Law Fund).

28. See note 27; S_1 is depicted more fully in chapter III, the transition process in chapter IV.

29. The detailed character of these upward avenues of influence is depicted in chapter III of the WOMP study.

5. The Twilight of Physical Descriptions and the Ascent of Normative Models

1. J. O. Wisdom, "Observations as the Building Blocks of Science in 20th Century Scientific Thought," *Boston Studies in the Philosophy of Science*, vol. VIII, R. Buck and R. Cohen, eds. (1972); see also J. O. Wisdom, "Scientific Theory: Empirical Content, Embodied Ontology and Weltanschauung," *Philosophy and Phenomenological Research*, 33 (1972).

2. See for instance, Carlos Castañeda, *Journey to Ixtlan, The Lessons of Don Juan* (1972), (on the Yaqui way of knowledge); or Gerardo Reichel-Dolnatoff, *Amazonian Cosmos* (1971).

3. Derek de Solla Price, *Little Science Big Science* (1965).

4. Viktor E. Frankl, "Reductionism and Nihilism," in *Beyond Reductionism*, A. Koestler and J. R. Smythies, eds. (1969).

5. Ralph Burhoe, "What Specifies the Values of the Man-Made Man?" *Zygon*, 6 (1971).

6. Michael Scriven, "Logical Positivism and the Behavioral Sciences," in *The Legacy of Logical Positivism*, P. L. Achinstein and S. Baker, eds. (1969).

7. Håkan Törnebohm, "United Studies," this volume.

8. Albert Wilson, "Systems Epistemology," this volume.

9. Albert Paolo Soleri, *Arcology: The City in the Image of Man* (1968); *Sketchbook* (1970); see also Henryk Skolimowski, "Paolo Soleri and the Philosophy of Urban Life," *Architectural Association Quarterly* (London, January 1971).

10. Ervin Laszlo, "Uses and Misuses of World System Models," this volume.

11. For further arguments see: Henryk Skolimowski, "Epistemology, The Mind and the Computer," in *Biology, History and Philosophy*, A. R. Breck and W. Yorgrau, eds. (1972).

6. Systems Epistemology

1. Lynn White, Jr., "The Historical Roots of Our Ecological Crises," *Science*, 155 (1967), pp. 1203–1207.

2. Robert Theobald, *Bulletin World Future Society*, 6 (1972), p. 5.

3. Roland Fischer, "A Cartography of the Ecstatic and Meditative States," *Science*, 174 (1971) pp. 897–904.

4. Alexis Carrel, *Man The Unknown* (New York: Harper & Bros. Publishers, 1935).

5. C. S. Lewis, *The Abolition of Man* (New York: The Macmillan Co., 1944).

6. Ervin Laszlo, "Systems Philosophy," *Main Currents in Modern Thought*, 28 (1971) pp. 55–60.

7. G. K. Zipf, *Human Behavior* (New York: Hafner Publishing Co., 1965).

8. Anatol Rapoport, "The Search for Simplicity," *The Relevance of General Systems Theory*, E. Laszlo, ed. (New York: Braziller, 1972).

9. Cf. Ervin Laszlo, *Introduction to Systems Philosophy*, chap. 11 (New York: Harper Torchbooks, 1973).

10. Mario Bunge, *Scientific Research* vol. 1, p. 51 (New York: Springer Verlag, 1967).

11. Alfred de Grazia, ed., *The Velikovsky Affair* (Hyde Park: University Books, 1966).

12. Fritz Zwicky, *Discovery, Invention, Research* (New York: The Macmillan Co., 1969).

13. Allen Hynek, *The UFO Experience* (Chicago: Henry Regnery Co., 1972).

8. *The World System and Human Values*

1. Jay W. Forrester, "Churches at the Transition between Growth and World Equilibrium," *Zygon* 7(3):145–67.

2. P. W. Bridgman, *The Way Things Are* (Cambridge, Mass.: Harvard University Press, 1959). Other physicists have found it necessary to delimit the duality of mind and matter. See, for instance, Erwin Schrödinger's *Mind and Matter* (London: Cambridge University Press, 1959). David Bohm has written, "Inanimate nature *is* both observer and observed," and also, "The observer is the totality of all that exists, and this is also the observed," p. 58 in his "Some Further Remarks on Order," in C. H. Waddington, ed., *Towards a Theoretical Biology*, vol. II (Chicago: Aldine Publishing Co., 1969).

3. For a view of life as a metastable pattern of flow or an open system, see A. Katchalsky, "Thermodynamics of Flow and Biological Organization," *Zygon* 6(2):99–125 (June 1971).

4. Ralph Wendell Burhoe, "Evolving Cybernetic Machinery and Human Values," *Zygon* 7(3):188–209 (September 1972).

5. Theodore Roszak, *Where the Wasteland Ends* (New York: Doubleday, 1972).

6. J. Z. Young, *A Model of the Brain* (London: Oxford University Press at the Clarendon Press, 1964).

7. For a fuller analysis, see Ralph Wendell Burhoe, "Five Steps in the Evolution of Man's Knowledge of Good and Evil," *Zygon* 2(1):77–96 (March 1967).

8. Alfred Emerson, "Dynamic Homeostasis," *Zygon* 3(2):129–68 (June 1968).

9. Erwin Schrödinger, *What Is Life?* (New York: Doubleday, 1956).

10. Norbert Wiener, *Cybernetics* (Boston: Houghton Mifflin, 1956); *Human Use of Human Beings: Cybernetics and Society* (Boston: Houghton Mifflin, 1950).

11. An interesting history of this is given in James D. Watson, *The Double Helix* (New York: Athenaeum, 1968).

12. J. Bronowski, "New Concepts in the Evolution of Complexity," *Zygon* 5(1):18–35 (1970).

13. A. Katchalsky, "Thermodynamics of Flow," (see n. 3).

14. Herbert A. Simon, *The Sciences of the Artificial* (Cambridge: MIT Press, 1960).

15. B.F. Skinner, "Phylogeny and Ontogeny of Behavior," *Science* 153 (1966):1205–13.

16. I use the term "ecological niche" in the sense Eugene Odum gives: "Ecologists use the term *habitat* to mean the place where an organism lives, and the term *ecological niche* to mean the role that the organism plays in the ecosystem: the habitat is the 'address,' so to speak, and the niche is the 'profession.'" Eugene P. Odum, *Ecology* (New York: Holt, Rinehart & Winston, 1963), p. 27.

17. See Holger Hydén, "Biochemical Changes Accompanying Learning," in Gardner Quarton, et al., *The Neurosciences* (New York: Rockefeller University Press, 1967).

18. Ralph Wendell Burhoe, "What Specifies the Values of the Man-Made Man," *Zygon,* 6(3):224–46 (1971).

19. Jay W. Forrester, *World Dynamics* (Cambridge, Mass.: Wright-Allen Press, 1971); D. H. Meadows, et al., *The Limits of Growth* (New York: Universe Books, 1972).

20. Burhoe, "Evolving Cybernetic Machinery," (see n. 4).

21. Herbert Simon, among others, has said it in his *Sciences of the Artificial* (see n. 14).

22. In *The Futurist,* 5(4):137–45 (August 1971), quoted from p. 144; italics mine.

23. *Zygon* 7(3):145–67 (September 1972), and in Dennis L. Meadows, ed., *Towards Global Equilibrium: Collected Papers* (Cambridge, Mass.: Wright-Allen Press, 1972).

24. For another systems dynamic view of religion as a key element in the norms or values for maintaining human society, see the chapter on "Energetic Basis for Religion" in Howard T. Odum, *Environment, Power, and Society* (New York: Wiley-Interscience, 1971).

25. Forrester, "Churches at the Transition," pp. 158–59.

26. Ralph Wendell Burhoe, "The Phenomenon of Religion Seen Scientifically," in Allan W. Eister, ed., *Changing Perspectives in the Scientific Study of Religion* (New York: Wiley-Interscience, in press).

27. Forrester, "Churches at the Transition," p. 159.

28. *Ibid.,* pp. 161–62.

29. *Ibid.,* p. 162.

30. *Ibid.*

31. *Ibid.,* p. 167.

32. *Ibid.,* p. 164.

33. *Ibid.,* p. 165.

34. See Donald T. Campbell, "On the Genetics of Altruism and the Counter-Hedonic Components in Human Culture," *Journal of Social Issues,* 29(3):1973.

35. See especially p. 144 in Ralph Wendell Burhoe, ed., *Science and Human Values in the 21st Century* (Philadelphia: Westminster Press, 1971).

Basic Bibliography

Abler, R., Adams, J., and Gould, P. *Spatial Organization*. Englewood Cliffs: Prentice-Hall, 1971.

Albertson, Peter, and Barnett, Margery. *Environment and Society in Transition*. (Annals of the New York Academy of Sciences, vol. 184), New York: The New York Academy of Sciences, 1971.

Almond, G. A., and Coleman, J. S., eds. *The Politics of the Developing Areas*. Princeton University Press, 1960.

Ashby, W. Ross, et al. *The Process of Model Building in the Behavioral Sciences*. Columbus: Ohio State University, 1970.

Bateson, Gregory. *Steps to an Ecology of Mind*. San Francisco: Chandler, 1972.

Bateson, Mary Catherine. *Our Own Metaphor*. New York: Knopf, 1972.

Berry, B. J. L., and Marble, D. F., eds. *Spatial Analysis*. Englewood Cliffs: Prentice-Hall, 1968.

Bertalanffy, Ludwig von, *General System Theory*. New York: George Braziller, 1968.

Black, Cyril E., and Falk, Richard A. *The Future of the International Legal Order*. Vols. I–IV. Princeton University Press, 1969–1972.

Brown, G. Spencer. *Laws of Form*. New York: Julian Press, 1972.

Brucan, S. *The Dissolution of Power*. New York: Knopf, 1971.

Bunge, Mario. *Scientific Research*. New York: Springer-Verlag, 1967.

Burton, J. W. *Systems, States, Diplomacy and Rules*. Cambridge University Press, 1968.

Carothers, J. Edward; Mead, Margaret; McCracken, Daniel D.; and Shinn, Roger L., eds. *To Love or to Perish*. New York: Friendship Press, 1972.

Chorley, R. J., and Haggett, P., eds. *Models in Geography*. London: Methuen, 1967.

Chorley, R. J., and Kennedy, B. A. *Physical Geography, A Systems Approach*. London: Prentice-Hall, 1971.

Churchman, C. West. *Design of Inquiring Systems*. New York: 1972.

Cohen, S. B., and Rosenthal, L. D. "A Geographical Model for Political Systems Analysis," *The Geographical Review*. LXI, no. 1 (1971), pp. 5–31.

Commoner, Barry. *Science and Survival*. New York: Viking Press, 1966.

——. *The Closing Circle*. New York: Knopf, 1971.

205

Dechert, C. R. "Integration and Change in Political and International Systems," *Positive Feedback: A General Systems Approach to Positive-Negative Feedback and Mutual Causality*. (J. H. Milsum, ed.) Oxford: Pergamon, 1968.

Deutsch, Karl. *Politics and Government*. Boston: Houghton Mifflin, 1970.

———. *Political Community at the International Level: Problems of Definition and Measurement*. Princeton University Press, 1953.

Easton, D. *A Framework for Political Analysis*. Englewood Cliffs: Prentice-Hall, 1965.

———. *A Systems Analysis of Political Life*. New York: John Wiley and Sons, 1965.

Falk, Richard A. *This Endangered Planet: Prospects and Proposals for Human Survival*. New York: Vintage, 1971.

———. *The Status of Law in International Society*. Princeton University Press, 1970.

Falk R., and Mendlovitz S., eds. *The Strategy of World Order*. World Law Fund, vols. I–IV, 1966–67.

Fisher, C. A., ed. *Essays in Political Geography*. London: Methuen, 1968.

Foerster, Heinz von, ed. *Cybernetics*. 5 vols. New York: Josiah Macy, Jr. Foundation, 1951–1955.

Forrester, Jay W. *Industrial Dynamics*. Cambridge, Mass.: The MIT Press, 1961.

———. *Principles of Systems*. (Preliminary Edition, 10 chaps.) Cambridge, Mass.: Wright-Allen Press, Inc., 1968.

———. *Urban Dynamics*. Cambridge, Mass.: Wright-Allen Press Inc., 1969.

———. *World Dynamics*. Cambridge, Mass.: Wright-Allen Press, Inc., 1971.

———. "Churches at the Transition between Growth and World Equilibrium," *Zygon* 7, No. 2 (Summer, 1972)

Frank, Lawrence K. *Nature and Human Nature: Man's New Image of Himself*. New Brunswick: Rutgers University Press, 1951.

Fuller, Buckminster. *Utopia or Oblivion*. Bantam Books, 1969.

Goody, B. *Perception of the Environment: An Introduction to the Literature*. University of Birmingham Center for Urban and Regional Studies, Occasional Paper No. 17, 1971.

Gorer, Geoffrey. *The Danger of Equality: Selected Essays*. London: Cresset, 1966.

Hanson, N. R. *Patterns of Discovery*. Cambridge, 1958.

Harvey, D. *Explanation in Geography*. London: Arnold, 1969.

Hawkins, David. *The Language of Nature*. San Francisco: W. H. Freeman, 1964.

Humphreys, W. C. *Anomalies and Scientific Theories*. London: 1965.

Huxley, Thomas H., and Huxley, Julian S. *Touchstone for Ethics*. New York: Harper, 1947.

Jackson, W. A. D., and Samuels, M. S., eds. *Politics and Geographic Relationships*. 2d ed. Englewood Cliffs: Prentice-Hall, 1971.

Kaplan, M. "The Systems Approach to International Politics," *New Approaches to International Relations*. New York: St. Martins, 1968.

Kasperson, R. G., and Minghi, J. V., eds. *The Structure of Political Geography*. Chicago: Aldine, 1969.

Keohane, R. O., and Nye, J. S., Jr., eds. *Transnational Relations and World Politics*. (*International Organization*, XXV, no. 3, summer 1971).

Kissinger, H. *Nuclear Weapons and Foreign Policy*. New York: W. W. Norton, 1969.

Klaus, G. *Kybernetik und Erkenntnistheorie*. Berlin, 1966.

Koestler, Arthur. *The Roots of Coincidence*. New York: Random House, 1972.

Kuhn, Thomas. *The Structure of Scientific Revolutions*. Chicago University Press, 1970.

Lakatos, I., and Musgrave, A., eds. *Criticism and the Growth of Knowledge*. Cambridge, 1970.

Laszlo, Ervin. *Introduction to Systems Philosophy*. New York: Gordon & Breach, 1972, Harper Torchbooks, 1973.

———, ed. *The Relevance of General Systems Theory*. New York: George Braziller, 1972.

———. *The Systems View of The World*. New York: George Braziller, 1972.

Laszlo, Ervin, and Wilbur, James, eds. *Human Values and Natural Science*. New York: Gordon & Breach, Science Publishers, 1971.

———, eds. *Human Values and the Mind of Man*. New York: Gordon & Breach, Science Publishers, 1972.

Lindberg, L. N., and Scheingold, S. A., eds. *Regional Integration: Theory and Research*. (*International Organization*, XXIV, No. 4, Autumn, 1970.)

Lorenz, Konrad. *On Aggression*. New York: Harcourt, Brace & World, 1966.

Maruyama, Magoroh. *1970 American Anthropological Association Cultural Futurology Symposium: Pre-Conference Volume*. University of Minnesota, Training Center For Community Programs, 1970. (mimeographed)

———. "The Second Cybernetics: Deviation-Amplifying Mutual Causal Processes," *Modern Systems Research for the Behavioral Scientist*. Edited by W. Buckley. Chicago: Aldine, 1968.

Matson, F. W. *The Broken Image: Man, Science and Society*. Garden City: Doubleday, 1966.

McClelland, C. A. "Applications of General Systems Theory," *International Politics and Foreign Policy*. Edited by J. Rosenau. New York: Free Press, 1961.

———. *Theory and the International System*. New York: Macmillan, 1966.

Mead, Margaret; Chapple, Eliot D.; and Brown, G. Gordon. "Report of the Committee on Ethics," *Human Organization*. 8, no. 2 (spring 1949) pp. 20–21.

Mead, Margaret. "Towards More Vivid Utopias," *Science*. 126, no. 3280 (November 8, 1957) pp. 957–961.

———. *Continuities in Cultural Evolution*. New Haven and London: Yale University Press, 1964. Reprinted in paperback, 1966.

———. "The Information Explosion," *The New York Times*. May 23, 1965, pp. 18–20.

207

————. "The Future as the Basis for Establishing a Shared Culture," *Daedalus* (Winter 1965) pp. 135–155.

————. "Museums in a Media-Saturated World," *Museum News.* 49, no. 1 (September 1970) pp. 23–25.

————. *Culture and Commitment.* Garden City, New York: Natural History Press/Doubleday, 1970.

————. "The Kalinga Prize," *Journal of World History (Unesco).* 13, no. 4 (1971) pp. 765–771.

Meadows, Dennis L. *Dynamics of Commodity Production Cycles.* Cambridge, Mass.: Wright-Allen Press, Inc., 1970.

Meadows, Dennis L., and Meadows, Donella H., eds. *Toward Global Equilibrium.* Cambridge, Mass.: Wright-Allen Press, Inc., 1973.

————, et. al., *The Limits to Growth.* New York: Universe Books, 1972.

Morgenthau, H. J. *Politics Among Nations.* 5th ed. New York: Knopf, 1972.

Nettl, P. "The Concept of System in Political Science," *Political Studies.* Vol. XIV, no. 3, October 1966.

Odum, Howard T. *Environment, Power and Society.* New York: Wiley-Interscience, 1971.

Olson, Richard. *Science as Metaphor.* Belmont, California, 1971.

Popper, K. *Conjectures and Refutations.* London, 1963.

————. *The Logic of Scientific Discovery.* London, 1959.

Ravetz, J. *Scientific Knowledge and Its Social Problems.* London, 1971.

Rosenau, J. *The Adaptation of National Societies: A Theory of Political System Behavior and Transformation.* New York: McCaleb-Seiler, 1970.

Saarinen, T. F. *Perception of Environment.* Washington, D.C.: Association of American Geographers, Resource Paper No. 5, 1969.

de Seversky, A. P. *Air Power: Key to Survival.* New York: Simon and Schuster, 1950.

Skolimowski, Henryk. "Science and the Modern Predicament," *New Scientist.* February 1972.

————. "Science in Crisis," *Cambridge Review,* January 1972.

————. "Problems of Rationality in Biology," in *Problems of Reduction in Biology.* Edited by F. Ayala, and Th. Dobzhansky. Macmillan, forthcoming.

————. "Rationality in Architecture," *R. I. B. A. Journal.* August 1972.

Soja, E. W. *The Political Organization of Space.* Washington, D.C.: Association of American Geographers, Resource Paper No. 8, 1971.

Sprout, H. and Sprout, M. "Environmental Factors in the Study of International Politics," *The Journal of Conflict Resolution.* Vol. I, no. 4 (1957) pp. 309–328.

————. *An Ecological Paradigm for the Study of International Relations.* Princeton University Press, 1968.

Spykman, N. J. *The Geography of Peace.* New York: Harcourt, Brace, 1944.

Stephens, J. "An Appraisal of Some System Approaches in the Study of International Systems," *International Studies Quarterly.* Vol. 16, no. 3, September 1972, pp. 321–349.

Taylor, A. M. "Evolution-Revolution, General Systems Theory, and Society," *Evolution-Revolution: Patterns of Development in Nature, Society, Man and Knowledge.* Edited by R. Gotesky and E. Laszlo. New York: Gordon and Breach, 1971.

Thomas, M. M., and Albrecht, Paul, eds. *World Conference on Church and Society, Official Report.* Geneva: World Council of Churches, 1967.

Tiselius, Arne, and Nilsson, Sam. *The Place of Value in a World of Facts.* Nobel Symposium 14. New York: Wiley-Interscience, 1967.

Törnebohm, H. *Reflections on Scientific Research. Scientia* (March-April, 1971).

————. "An essay on theory of research with special reference to physics, II." *Reports from the department of theory of science,* no. 36, Göteborg, 1972.

————, and Radnitzky, G. *Forschung als innovatives System. Zeitschrift für allgemeine Wissenschaftstheorie.* Vol. 2, no. 2, 1971.

Toulmin, Stephen. *Human Understanding.* Princeton University Press, 1972.

Waddington, C. H., ed. *Biology and the History of the Future.* Edinburgh: Edinburgh University Press, 1972.

————. *The Ethical Animal.* London: Allen and Unwin, 1960; Atheneum, 1961.

Ward, Barbara, and Dubos, René. *Only One Earth.* New York: Norton, 1972.

Weaver, Warren. "Science and People," *Science.* 122, no. 3183 (December 1955) pp. 1255–1259.

Weiss, Paul A., ed. *Hierarchically Organized Systems in Theory and Practice.* New York: Hafner Publishing Co., 1971.

Whitehead, A. N. *Science and the Modern World.* New York: New American Library, 1948.

Wiener, Norbert. *Human Use of Human Beings.* Boston: Houghton Mifflin, 1954.

Wright, Q. *A Study of War.* University of Chicago Press, 1964.

Notes on Contributors

RALPH WENDELL BURHOE was for most of his distinguished career the executive officer of the American Academy of Arts and Sciences. He is currently Research Professor in Theology and Science, Meadville/Lombard Theological School; editor of *Zygon, Journal of Religion and Science;* and secretary of the Center for Advanced Study in Religion and Science.

RICHARD A. FALK is Albert G. Milbank Professor of International Law and Practice at Princeton University. He is a vice president of the American Society of International Law, a member of the board of directors of the Fund for Peace and the Foreign Policy Association, and is currently the director of the North American section of the World Order Models Project of the World Law Fund. His most recent book is *This Endangered Planet* (Random House, 1971).

JAY W. FORRESTER is Germeshausen Professor at the Massachusetts Institute of Technology. He is the recipient of several honors in America and abroad, including the Medal of Honor of the Institute of Electrical and Electronic Engineers, and the Award for Outstanding Accomplishment by the Systems, Man and Cybernetics Society. His numerous writings include *Principles of Systems* (Wright-Allen Press, 1968) and *World Dynamics* (Wright-Allen Press, 1971).

ERVIN LASZLO is professor of philosophy at the State University of New York at Geneseo and currently Visiting Research Fellow at the Center of International Studies at Princeton University. He serves as editor of the International Library of Systems Theory and Philosophy, and chairman of the North-East Division of the Society for General Systems Research. His books in the area of systems philosophy include *The Systems View of the World* (Braziller, 1972) and *Introduction to Systems Philosophy* (Gordon and Breach, 1972; 2d ed. Harper Torchbooks, 1973).

MARGARET MEAD is past president of the Society for General Systems Research as well as of the American Anthropological Association. She is the recipient of numerous awards, including the Viking Medal in General Anthropology of the Wenner Gren Foundation. Currently Dr. Mead is

211

Curator Emeritus of Ethnology at the American Museum of Natural History. Her most recent book is *Blackberry Winter,* a story of her early years.

HENRYK SKOLIMOWSKI is professor of philosophy in the Department of Humanities at the University of Michigan's College of Engineering. He has been a visiting professor in numerous institutions, including the Warsaw Institute of Technology, the Hebrew University of Jerusalem, and the School of Architecture in London. His writings have appeared in journals of philosophy, science, and architecture both in America and in Europe.

ALASTAIR M. TAYLOR is professor of political science as well as of geography at Queen's University in Kingston, Ontario. He has been visiting professor at the Universities of Edinburgh and the West Indies, and has served in addition in the secretariats of UNRRA and the United Nations. He is the coauthor of *Civilizations: Past and Present,* and pioneered the application of general systems theory to societal processes on national and international levels in contributions to books such as *Evolution-Revolution: Patterns of Development in Nature, Society, Man and Knowledge,* and *Integrative Principles of Modern Thought.*

HÅKAN TÖRNEBOHM is professor of theory of science at the Institute for Theory of Science at the University of Gothenburg in Sweden, and serves as director of the institute. He has also served as professor at the University of Khartoum in Sudan. Dr. Törnebohm's numerous contributions have been in the fields of philosophy of science, especially theory of relativity, theory of scientific research, and the application of general systems theory to scientific research in the social context.

ALBERT WILSON heads his own research and consulting firm in California. He is also teaching courses on the subjects of technology, business, and futurology at the University of California at Los Angeles. His primary background is in mathematics and astronomy but has increasingly shifted his interest toward the fields of epistemology and values. He is the author of *New Methods of Procedure* (Springer-Verlag, 1967) and the editor, with Lancelot Law Whyte and his wife Donna Wilson, of *Hierarchical Structures* (American Elsevier, 1969). He is currently finishing a book with Donna Wilson, *Futures: The Dynamics of Normative Systems.*

212

INDEX